Do the Inhabitants of other Planets ever visit this Earth?

I propose in this connection to make a few remarks on the following: Mr. Henry Wallace and other persons of Jay, Ohio, have recently detailed to me the annexed. There are thousands of such cases on record. These gentlemen state, that sometime since on a clear and bright day, a shadow was thrown over the place where they were; this necessarily attracted their attention to the Heavens, where they one and all beheld a large and curiously constructed vessel, not over one hundred yards from the earth. They could plainly discern a large number of people on board of her, whose average height appeared to be about twelve feet. The vessel was evidently worked by wheels and other mechanical appendages, all of which worked with a precision and a degree of beauty never yet attained by any mechanical skill upon this planet.

And that this ship which Mr. Wallace and others saw, was a vessel from Venus, Mercury, or the planet Mars, on a visit of pleasure or exploration, or some other cause; I myself, with the evidence at hand, that I can bring to bear on it, have no more doubt of, than I have of the fact of my own existence. This, mind, was no phantom that disappeared in a twinkling, as all phantoms do disappear, but this aerial ship was guided, propelled and steered through the atmosphere with the most scientific system and regularity, at about six miles an hour, though, doubtless, from the appearance of her machinery, she was capable of going thousands of miles an hour, and who knows but ten thousand miles an hour...

Example of Airship c.1896.

THE REAL COWBOYS & ALIENS
EARLY AMERICAN UFOS
1800-1864

By Noe Torres & John LeMay

ROSWELL BOOKS.COM
Roswell, New Mexico · Edinburg, Texas

© COPYRIGHT 2019 Noe Torres & John LeMay
All rights reserved.

Cover Illustration by Christopher Martinez
https://bit.ly/2PivOS6

Torres, Noe.
LeMay, John.
 The Real Cowboys and Aliens:
 Early American UFOs (1800-1864)
  1. History—Old West. 2. Ufology—Study of
     Unidentified Flying Objects. 3. Folklore, Old
     West.

In memory of our friend,
Elvis E. Fleming, the Roswell historian who first
used the title "Cowboys and Aliens" for a 2005
essay on the history of Roswell, NM.

ACKNOWLEDGEMENTS

The authors would like to acknowledge the wonderful assistance received from our friends, Jacques Vallee, world-renowned UFO researcher and author, and Ruben Uriarte, another world class UFO researcher who has spent years investigating the phenomenon for the Mutual UFO Network. Also, we would like to thank paranormal researcher Jerome Clark, whose many publications on unexplained phenomena dating back to early human history served as a great inspiration to us in the writing of this book.

PREFACE

IT HAS NOW BEEN ten years since the authors wrote their book *The Real Cowboys and Aliens: UFO Encounters of the Old West*. The idea for the book came about from the hype surrounding the 2011 summer blockbuster film *Cowboys and Aliens*. In that film, based upon a comic book, actors Daniel Craig and Harrison Ford played gunslingers that faced off against extraterrestrial "bad guys," intent on stealing the Earth's rich deposits of underground gold.

At the time of the movie's release, the authors of this book felt the public would be intrigued to know that there were indeed UFO and alien encounters during the era of the Old West, resulting in our first book, which was aimed primarily at readers in middle school and high school.

To our surprise, not only did the younger readers love our book, but so did the adults! The response was so extremely positive that the idea of another book that expanded the topic and was aimed more directly at adults immediately came to mind. There was one problem, however. At the time we felt, somewhat naively, that we had covered most of the significant UFO encounters from the 1800s and that we would find few other stories to include in a follow-up book. Boy, were we wrong!

In the years following the publication of our first book, many more newspapers and magazines from the 1800s were digitized and became accessible using advanced Boolean search technologies. We suddenly had in our hands a multitude of "new" stories of bizarre encounters from the 19th century and early 20th century. As we enthusiastically uncovered hundreds of strange cases from America's early decades, we faced the stark realization that our first book had barely scratched the surface.

What is most remarkable, perhaps, is that all but one of the tales in this new book take place before the first chapter of our original book! We began our 2011 book with a reported 1864 UFO crash in the Rocky Mountains and ended it with a chapter on alien cattle rustlers from 1897. In a little over 100 pages, we covered what we felt at the time were most of the more interesting early American encounters, but as stated, we were completely wrong.

Before we undertake our return journey to the 1800s, some context regarding the time period will be helpful for most readers. For starters, some readers may wonder why this book, covering UFO sightings from 1800 to 1864, is part of a series titled *The Real Cowboys and Aliens*, when America's

"Old West" period technically did not begin until 1865. However, in truth, America's expansion to the West from the original settlements along the East coast began much earlier than 1865. Historians place the beginning of America's Western Expansion in the year 1803, with the acquisition from France of a vast section of land known as the "Louisiana Purchase."

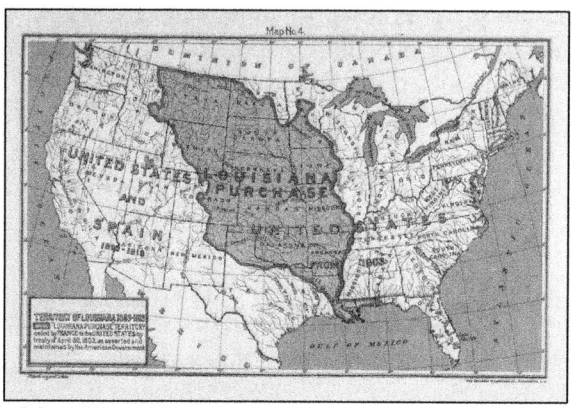

Map Shows Extent of the Louisiana Purchase Area

America's movement to the West is said to have taken place in four phases: the Louisiana Purchase, the Gold Rush, the Oregon Trail, and then the time of "Manifest Destiny," a belief that the expansion of the United States across the American continent was literally God's destiny at work.

Before the Louisiana Purchase, the area west of the Mississippi River had been primarily controlled by France. Then the French Indian War broke out in 1754 and went on until 1763. Afterward, France sold its vast land holdings west of the Mississippi to the United States in what was known as the Louisiana Purchase, expanding the U.S. by 828,000 square miles for a mere $15 million. Thus, a huge land mass west of the Mississippi officially became part of the United States in 1803.

After the Louisiana Purchase, the Westward migration of new settlers followed almost immediately, and by 1804 a territorial government was established. To explore these vast new lands, Thomas Jefferson commissioned Meriwether Lewis to mount a thorough exploration of the new territories. Lewis then chose William Clark to be his partner in the excursion, thus creating the historic Lewis and Clark Expedition. The resulting 8,000-mile journey took more than two years and was a great success, traversing and mapping out previously uncharted areas of North America.

America's next great step in the settling of the Western territory came about with the establishment of the Oregon Trail, a roughly 2,000-mile path from Independence, Missouri, to Oregon City, Oregon. By 1843 a vast procession moved West that included 120 wagons, 1,000 people and thousands of livestock. From that point forward, there would be many pioneers travelling the Oregon Trail by wagon, until the arrival of the railroad in 1869, after which the trains replaced wagons.

The next phase of America's Western migration was the California Gold Rush. The gold rush was sparked in 1848 when gold nuggets were found in the Sacramento Valley. The discovery just happened to coincide rather perfectly with another huge historical event, the signing of the Treaty of Guadalupe Hidalgo that ended the Mexican-American War. The short war lasted from 1846 to 1848, as the U.S. and Mexico fought over the control of California, among other states. With California now in the possession of the United States as of 1848, coupled with the news that gold had been discovered there, a massive western migration followed. Within a year, California's Anglo population exploded from about 1,000 settlers to 100,000!

Panning for Gold in California, c. 1850

By the 1850s, debates began to flare up as to the moral ethics of slavery, with the North favoring its abolition and the South favoring its retention. War officially broke out when the newly formed Confederate Army attacked Fort Sumter in South Carolina on April 12, 1861. By the time hostilities ended four years later, the war claimed between 620,000 and 750,000 lives.

The Confederacy collapsed, the slaves were freed, and the Reconstruction era began. And with that also began what is called the "Wild West," which had its share of interesting UFO sightings that will be the subject of our next book in this series, covering the time period 1865-1895.

We hope that this very brief refresher on history will help readers to truly understand the unusual incidents related in this book within the context of the time period in which they occurred. When you are done reading this book, we hope that it will have given you a very different perspective on American history, one filled with strange "Signs and Wonders" in the skies, sightings that sound very much like modern day UFO cases, tales of mysterious "Little People" living in the wild, stories of abductions similar to modern day tales of alien abduction, and the strange phenomena of slimy goo that reportedly fell from outer space, called "star rot."

One final note about the time period in which these stories take place. We should commence this journey by understanding what the early American settlers thought about the possibility of extraterrestrial life elsewhere in the Universe-- the answer being that they were so preoccupied with the harsh conditions of daily life on the frontier that they rarely, if ever, thought about it. Perhaps they occasionally thought about life on other worlds as a jest, tall tale, or in the context of religion. There was certainly little scientific thought about the topic, with a few notable exceptions, such as Benjamin Franklin, Thomas Jefferson, and several other important thinkers of the time.

Also, it is important to know that the people of the 1800s did not have the terminology we take for granted today, including words such as unidentified flying objects, rocket engines, spaceships, jet planes, extraterrestrials, space travel, etc. If they ever thought about traveling between worlds, the pioneers would view it in allegorical or religious terms, not in scientific terms.

In their dimness of understanding, any object that was seen moving rapidly across the sky was described only by the words and images with which they were familiar, including terms such as shooting star, meteor, comet, angel, dragon, serpent, etc. Also, for example, if a person from the 1800s had encountered a four-foot-tall grey alien wandering in the forest, the creature might have been called a "fairy," a "dwarf," a "wild man," or a "strange beast."

It is true that there are not many true "cowboys" in this volume of *The Real Cowboys & Aliens* series and that many of our stories take place not in the West but along the East coast of the United States. Nevertheless, the incidents were all experienced by normal, everyday citizens of early America. Like the cowboys, these pioneers, regardless of where they lived, rode on horses, shot guns and frequented saloons from time to time. Many of the familiar elements of the "Old West" were as much a part of life in the East as they were in the towns of the West.

The book you now hold in your hands, or on your electronic device, is well over 200 pages, double the length of our first book! And these new stories are definitely not "leftovers" from our previous book; on the contrary, some of these newly discovered stories are actually even more bizarre and fascinating than the stories we included in our first book.

For instance, would you have ever guessed that there was an incident in the 1800s that mirrored the classic 1958 horror film *The Blob*? Or, that the "Men in Black" were seen as far back as the Civil War? What about a sword that fell from the heavens down to Earth? Or, a gigantic clam-like unidentified submersible object (USO) seen in the Atlantic Ocean? Going back even earlier in time, did you know that explorers Lewis and Clark may have uncovered evidence of a tribe of extraterrestrials in their travels? Or, that Thomas Jefferson once investigated a UFO sighting?

Whereas our original little book was the birth of an idea, this book and the next two in the series are the culmination of that idea grown to full maturity. As we examine the most amazing UFO cases from early in America's history, prepare to be astonished!

CONTENTS
ACKNOWLEDGMENTS vi
PREFACE vii

CHAPTERS

1.	Thomas Jefferson and the UFOs	17
2.	Aliens in Alaska	29
3.	Lewis and Clark and Little Men from Mars	39
4.	UFOs, Crop Circles and Fairies	49
5.	The Flying Humanoids	59
6.	The School Teacher and the UFO	69
7.	Floating UFO of the Atlantic	77
8.	The UFO Looked Like a Turtle	85
9.	Slime from the Stars	93
10.	Encounter at Mount Adams	103
11.	The UFO Before Kecksburg	111
12.	Night of the Leonids	112
13.	Little Green Medicine Men	127
14.	Return of the Space Slime	133
15.	That's No Moon!	141
16.	Zapped by a UFO	149
17.	UFOs Over Campus	155
18.	The 18-Inch Humanoid	163
19.	Huge UFO in Ohio	171
20.	Night of the UFOs	179
21.	UFOs Over the Missouri River	189
22.	Spaceships Over the Old South	197
23.	Mysterious Airship Over New York	203
24.	Houston, We Have Giant Aliens!	209
25.	Men in Black in the Civil War	217
26.	Army from Outer Space	227
27.	The Rocky Mountain Meteor	235
28.	The Blob From Outer Space	241
29.	Mummy From Mars	245

INDEX 254
ABOUT THE AUTHORS 258

THOMAS JEFFERSON AND THE UFOS

April 5, 1800
Baton Rouge, Louisiana

ALTHOUGH NORTH AMERICA's "Old West" period (1865-1895) was still decades away, the start of the 1800s featured several very interesting cases concerning unidentified flying objects (UFOs), including two that involved one of America's founding fathers, Thomas Jefferson, who served as U.S. vice-president (1797-1800) and then president (1801-1809). When the first of these "UFO" cases occurred, in 1800, Jefferson was also president of the American Philosophical Society (APS). Established in 1743 by Benjamin Franklin and others, the APS was America's first organization entirely devoted to learning and

17

scholarship. Its membership included some of the period's brightest minds, and many scholars argue that members were particularly interested in the subject of extraterrestrial life.

On June 30, 1880, Jefferson received a strange communication from George Dunbar, a plantation owner and amateur astronomer in Natchez, Louisiana, containing what may be America's first documented UFO sighting. Dunbar, known for his engineering and scientific talents, was so highly interested in astronomy that he built an observatory in 1799 near his home in Natchez. Regarding the observation of celestial events, Dunbar was certainly not disinterested or untrained.

"A phenomenon was seen to pass Baton Rouge on the night of the 5th of April 1880," Dunbar begins in his letter to Jefferson. He explains that a number of witnesses saw a "wholly luminous" flying object "the size of a large house" moving incredibly fast above them, at an altitude of about 200 feet. As the object, crimson red in color, passed overhead, nighttime turned into daylight for the group of spectators observing its flight. Around the area where they were standing, they felt "the effect of sun-beams," but if they looked away from their immediate area, they saw the darkness and the stars in the night sky. In addition to the beams of light that briefly flooded them as the object passed, the spectators also felt "a considerable heat" but no sensation of an electrical charge in the area.

EARLY AMERICAN UFOS: 1800-1864

> [25]
>
> ---
>
> No. III.
>
> *Description of a singular Phenomenon seen at Baton Rouge, by William Dunbar, Esq. communicated by Thomas Jefferson, President A. P. S.*
>
> NATCHEZ, June 30th, 1800.
>
> Read 16th January 1801.
>
> A PHENOMENON was seen to pass Baton Rouge on the night of the 5th April 1800, of which the following is the best description I have been able to obtain.
>
> It was first seen in the South West, and moved so rapidly, passing over the heads of the spectators, as to disappear in the North East in about a quarter of a minute.
>
> It appeared to be of the size of a large house, 70 or 80 feet long and of a form nearly resembling Fig. 5. in Plate, iv.
>
> It appeared to be about 200 yards above the surface of the earth, wholly luminous, but not emitting sparks; of a colour resembling the sun near the horizon in a cold frosty evening, which may be called a crimson red. When passing right over the heads of the spectators, the light on the surface of the earth, was little short of the effect of sun-beams, though at the same time, looking another way, the stars were visible, which appears to be a confirmation of the opinion formed of its moderate elevation. In passing, a considerable degree of heat was felt but no electric sensation. Immediately after it disappeared in the North East, a violent rushing noise was heard, as if the phenomenon was bearing down the forest before it, and in a few seconds a tremendous crash was heard similar to that of the largest piece of ordnance, causing a very sensible earthquake.
>
> I have been informed, that search has been made in the place where the burning body fell, and that a considerable portion of the surface of the earth was found broken up, and every vegetable body burned or greatly scorched. I have not yet received answers to a number of queries I have sent on, which may perhaps bring to light more particulars.
>
> F

The Original Report from Thomas Jefferson

Travelling rapidly from the southwest to the northeast, the object was above them only a matter of seconds before it could be seen in diminishing size off in the distance. Then suddenly, the area around the spectators was blasted by what was presumably the object's sonic boom. "Immediately after it disappeared in the North East, a violent rushing sound was heard," Dunbar wrote, "as if the phenomenon was bearing down the forest before it, and in a few seconds a tremendous crash was heard similar to that of the largest piece of ordnance, causing a very sensible earthquake."

Portrait of Thomas Jefferson (Public domain)

The assumption was that the object had crashed, although Dunbar expresses his frustration at not receiving any word on whether the object itself was found. "A considerable portion of the surface of the earth was found broken up," he wrote, "and every vegetable body burned or greatly scorched." No mention has ever been made of the recovery of

EARLY AMERICAN UFOS: 1800-1864

any fragments of the object itself. Could this have been a landing rather than a crash?

Historian Bill Thayer does not believe the object was a meteor, saying, "... if the observation of the object's size is anywhere near accurate, it was not a meteorite; an object of this size, entering earth's atmosphere at a speed typical of objects falling to earth from space, would probably have left a much larger trace of itself, and would almost certainly have killed the observer and anyone else near the fall. Scientists currently gauge the size of the iron meteor that created Arizona's Meteor Crater, for example, at roughly 50 meters, only about twice the estimate reported by Dunbar."

NASA Meteor Image

Thayer has also taken Dunbar's description of the incident and calculated that the object was moving at a speed of approximately 2,200 kilometers per hour (1,367 mph), far below the minimum speed of a meteor freefalling to Earth from space, which would be about 11,000 kilometers per hour (6,835

THE REAL COWBOYS & ALIENS

mph). Thayer concludes, "... if it was a house-sized object coming in at a meteoric speed, it would have been a huge event, with no survivors for miles, flattened trees, etc."[1]

In the June 14, 1968 issue of *Science* magazine, William T. Powers suggests that although most historians and scientists consider that the Dunbar account describes the descent of a meteor, he is not sure. "We cannot yet be certain that Dunbar's object was in fact a meteor," Powers writes. The author seems especially impressed that the object was rectangular in shape, according to a sketch drawn by Dunbar, and was not rounded like most meteors are, having come through the incredible heat and pressure of the Earth's atmosphere.

Unfortunately, this is all that is known about what may have been the first documented sighting of a UFO in the newly minted United States of America. Even less is known about Thomas Jefferson's second brush with a strange visitor from outer space, thirteen years later, in 1813, which will be covered more extensively in a later chapter.

Four years after leaving the White House, Jefferson received a strange communication regarding a possible UFO sighting from two gentlemen identifying themselves as Edward Hansford and John L. Clarke, saying that they "earnestly solicit that your honor will give us your opinion on the following extraordinary case." Especially unusual was the fact that the UFO seemed to shape-shift several times as it moved

---

[1] http://penelope.uchicago.edu/Thayer

EARLY AMERICAN UFOS: 1800-1864

across the sky, sometimes appearing to be turtle-shaped and later becoming more elongated like a "human skeleton."

Portion of the Letter from Hansford - Clarke

In their letter from Portsmouth, Virginia, dated July 31, 1813, the men write, "At hour on the night of the 25th instant, we saw in the South a Ball of fire full as large as the sun at [meridian] which was frequently obscured within the space of ten minutes by a smoke emitted from its own body, but ultimately retained its [brilliancy], and form during that period, but with apparent agitation."

Historians have identified Edward Hansford as the keeper of the Washington Tavern in the Town of Portsmouth, Virginia. Hansford later became harbormaster for the Norfolk and Portsmouth district. John L. Clarke, a native of Newport, Rhode Island, had recently been discharged as a

THE REAL COWBOYS & ALIENS

master or midshipman in the United States Navy. He later became a sea captain in Baltimore, where he was residing at the time of his death.

The letter about their strange sighting is held today in the papers of Thomas Jefferson at the Library of Congress, designated as "Edward Hansford to Thomas Jefferson. 07-31, 1813, available online at the time of this writing at: www.loc.gov/item/mtjbib020964/.

Scholars categorize the topic of the letter as "An Atmospheric Phenomenon." No evidence exists in documents that Jefferson ever responded to the letter in any manner or that he took any sort of action to investigate the phenomenon described.

The secret society known as the Freemasons, with whom Jefferson frequently associated, has long been suspected of having a strong interest in the topic of extraterrestrial life. Although no documents exist that prove Jefferson was ever a member of the Freemasons, he was certainly surrounded, and very likely heavily influenced, by members of the secret society, including George Washington, John Hancock, Benjamin Franklin, and Paul Revere. Many members of the American Philosophical Society, including Franklin, were high order Freemasons. In the Library of Congress is a painting of Washington dressed in Freemason garb and surrounded by symbols of Freemasonry, including what looks like a flying saucer hovering in the sky above him.[2]

---

[2] See https://www.loc.gov/pictures/item/96518222/.

EARLY AMERICAN UFOS: 1800-1864

Long a subject of conspiracy theories, the Freemasons are often linked with an elite global movement known as the *Illuminati*, who are said to secretly control important human affairs and possibly have links to extraterrestrials.

George Washington as a Freemason (Library of Congress)

Freemason Benjamin Franklin, America's greatest early inventor and philosopher, was a futurist who predicted anti-gravity transportation, a technology often observed in UFOs. He wrote, "The rapid progress true science now makes, occasions my regretting sometimes that I was born so soon: it is impossible to imagine the height to which may be carried, in a thousand years, the power of man over matter; we may perhaps learn to deprive large masses of their gravity, and give them absolute levity for the sake of easy transport."[3]

Portrait of Ben Franklin, Freemason
(Library of Congress)

Franklin also spoke about the inhabitants of other worlds, writing, "I believe that Man is not the

---

[3] Letter to Dr. Priestley, Feb. 8, 1780.

EARLY AMERICAN UFOS: 1800-1864

most perfect Being but One, rather that as there are many Degrees of Beings his Inferiors, so there are many Degrees of Beings superior to him."[4] In the same document, Franklin adds, "... when I stretch my Imagination thro' and beyond our System of Planets, beyond the visible fixed Stars themselves, into that Space that is every Way infinite, and conceive it filled with Suns like ours, each with a Chorus of Worlds forever moving round him, then this little Ball on which we move, seems, even in my narrow Imagination, to be almost Nothing ..."

An episode of the television series *Ancient Aliens* titled "Aliens and the Founding Fathers" (Season 3, Episode 11) claimed that Franklin and Jefferson were more than just mildly interested in extraterrestrials. They both believed in the "plurality of worlds" and were fascinated by the theories of astronomer William Herschel (1738-1822). Known as the discoverer of the planet Uranus, Herschel was a strong believer in extraterrestrial life and thought it possible that life existed on the other planets in our solar system and on the moon. In "Aliens and the Founding Fathers," author David Childress points out that Franklin and Jefferson were members of a secret society of intellectuals known as the Lunar Society, which would meet to discuss topics such as Herschel's ideas about life on other worlds.

As someone who was no doubt influenced by philosophical arguments in favor of extraterrestrial life, could Thomas Jefferson have had an unusually

---

[4] *Articles of Belief and Acts of Religion*, November 20, 1728.

strong interest in extraterrestrials? If so, these two early UFO incidents may have held a special significance to him and to other thinkers of the early 1800s.

ALIENS IN ALASKA
Circa 1800, Koyuk, Alaska

AT THE START OF THE 1800s, Alaska was Russian territory, was still a barren wilderness, and was chiefly populated by native Alaskans, often referred to as "Eskimos." Russia never established many settlements in Alaska, and the population of Russians in Alaska never exceeded 700. It wasn't until Alaska was purchased by the United States in 1865 and Americans began making the long trek north in search of gold that stories were first heard from the natives about flying objects and strange beings. One of the most striking of these stories comes from the village of Koyuk, located in the west central part of Alaska, near the city of Nome.

THE REAL COWBOYS & ALIENS

The story was first published on October 17, 1988, by Janet Ahmasuk (1943-2012), who was a reporter for the *Nome Nugget*, Alaska's oldest newspaper. For her article, Ahmasuk interviewed Lois Foster, whose family originated in Koyuk and whose grandmother Gadayroak told of a "silvery looking disk that sailed through the air" and of the "little men" who came from the sky and settled in Koyuk after their spaceship became disabled.

"Little People have appeared and disappeared a great many times in many different villages throughout this part of Alaska," Ahmasuk wrote, "Sometimes they stay awhile, and sometimes their appearance is so brief and bizarre that folks wonder about their own sanity.... According to Lois Foster, whose family originates in Koyuk, her great-grandmother told of three little men who came to the village on a silvery looking disc that sailed through the air."

The small humanoids were described as very strong, being able to pick up and carry large logs without any assistance. According to native folklore, these little people, referred to as *isrizaq* or *invaqalik*, should never be attacked by humans, because to do so would bring great harm to the humans, due to the creatures' vastly superior strength. To illustrate this point, Ashmasuk told a story of a local reindeer herder who thought he saw the corpse of a large deer suddenly begin to move by itself, only to discover that it was being carried by an incredibly strong "little man," standing underneath the corpse.

EARLY AMERICAN UFOS: 1800-1864

Little People visit the village

Portion of Original Article from Nome Nugget, 10-27-1988

The three strange visitors lived among the natives of Koyuk for many years, becoming fluent in their language and adapting in every way they could. Since they were all of the male gender, they could not reproduce and eventually all died without leaving heirs, according to Lois Foster's grandmother, Gadayroak.

*Reporter Janet Ashmasuk
(Nome Nugget 6-7-2012)*

THE REAL COWBOYS & ALIENS

Bob and Lois Foster, left, show new owners Janis and Larry Kile, right, just how many items are listed in the Sears catalog.

Lois Foster (2nd from left) in
Quad City Herald, 5-31-79

In many conversations with the beings over the years, the villagers learned that the "little men" had remained in Koyuk because "a mechanism had broken down in their craft, permanently disabling it." The strangers expressed that they had given up any hope of ever returning to their home.

Lois' grandmother met the three unusual men when she was a very young girl around the year 1910. The three "little men" were already extremely elderly at the time but were still alive. It is estimated that the creatures passed away sometime after that, having lived on earth for well over 100 years.

EARLY AMERICAN UFOS: 1800-1864

The story of the three small humanoids that crash landed near Koyuk around 1805 first surfaced in 1988 after a number of similar, strange apparitions were reported to local authorities in the vicinity of Nome. Mark Chorvinsky wrote in *Fate* magazine (January 1990, pages 19-28): "During the week of August 23rd, 1988, there was a number of sightings of little green men in Nome, Alaska. The entities were observed at close range over a period of five separate nights by more than a dozen local citizens. The little green men allegedly glowed, were very speedy, could change color, and were able to be run over by a car without harm. Eskimo lore is filled with tales of little men, some going back hundreds of years, and the 1988 sightings caused locals to wonder whether the entities were the little men Eskimos call *isrizaq* or *invaqalik*. *Nome Nugget* reporter Janet Ahmasuk looked into the case and during her investigation collected numerous accounts of the Alaskan little people, past and present."

Chorvinsky quotes Ahmasuk as saying, "I have heard little green men stories for as long as I've lived here from reindeer herders, miners, highway camp workers, village folks, folks that have lived here for a long time and folks that just moved here and wouldn't know an *isrizaq* or an *invaqalik* if they shook hands with him. Stories come from knowledgeable people and also folks that are in no shape to be knowledgeable. They seem to be part of the as yet unexplained phenomena of this area." Nome in 1988 may have nothing to do with the 1805 incident at Koyuk, but the reporter involved

THE REAL COWBOYS & ALIENS

> **LITTLE GREEN MEN IN ALASKA**
> During the week of August 23rd, 1988, there was a number of fascinating sightings of little green men in Nome, Alaska. The entities were observed at close range over a period of five separate nights by more than a dozen local citizens. The little men allegedly glowed, were very speedy, could change color, and were able to be run over by a car without harm. Eskimo lore is filled with tales of little men, some going back hundreds of years, and the 1988 sightings caused locals to wonder whether the entities were the little men Eskimos call "isrizaqs" or "invaqalik." Nome *Nugget* reporter Janet Ahmasuk looked into the case and during her investigation collected numerous accounts of the Alaskan little people, past and present. She writes: "I have heard little green men stories for as long as I've lived here from reindeer herders, miners, highway camp workers, village folks, folks that have lived here for a long time and folks that just moved here and wouldn't know an isrizaq or an invaqalik if they shook hands with him.
>
> *Fate magazine, Jan. 1990, pp. 19-28.*

in investigating both stories seemed to think there might have been a connection. In the October 6, 1988 issue of the *Nome Nugget*, Ahmasuk reported on the 1988 apparitions of "little green men" in Nome. A group of teenagers first encountered one of the strange creatures while out on a drive after work. Little did they realize that they would soon experience something that would haunt them for the rest of their lives. At about 2 a.m., the teenagers noticed a strange green glow in their rearview mirror. The driver stopped, pulled over, and turned the car around to look for the source of the eerie light. Deciding to have a closer look, the young man drove back onto the road, this time in the direction of the light.

Their car soon approached the figure of a small humanoid walking in the center of the road. The headlights first disclosed legs and feet in motion. As they drew closer, they were shocked to see, a creature no more than three feet tall whose entire

EARLY AMERICAN UFOS: 1800-1864

body glowed with a "greenish luminescence" and seemed to be semi-transparent.

According to the witnesses, the strange being "appeared somewhat transparent sort of like a hologram," and yet it gave the impression of a well-developed male humanoid, appearing to have a chiseled physique, broad shoulders, and muscular legs that resembled those of a trained athlete.

As the "little green man" became aware of the vehicle right behind it, it began to run extremely fast, with the car in hot pursuit, reaching speeds of nearly 50 miles an hour. Coming up right behind the runner, the teenage driver attempted to swerve around the creature but was not successful and struck the humanoid, running it over with the car. What happened next shocked and terrified the teenagers beyond all measure.

Instead of feeling the bump from having run over the being, the occupants of the car felt absolutely nothing. The strange creature "seemed to flatten out" and no harm came to it at all from being struck by the speeding car. Having witnessed this incredible event, the confused and alarmed teenagers left the area and rushed into town to find other people that could corroborate their story.

After picking up several friends and co-workers, they returned to the scene of their encounter, where they soon found the "little green man" standing beside the road. As they all gaped at the sight, the creature's luminescent body changed from green to silver. For the first time, the teens also noticed it had bright red eyes and was emitting a sound like a "dry, whistling hiss."

THE REAL COWBOYS & ALIENS

Two of the more daring of the teenagers got out of the car and attempted to chase the creature, but it quickly turned on them and chased them back into the car. Scared out of their wits, the teens headed back to town, leaving the creature alone, standing alongside the road.

The encounter was repeated the following night, on August 25, at approximately the same location and at about the same time. Hoping to further investigate the previous night's sighting, the same group of teenagers returned to the scene, bringing along with them several other townspeople, including the station manager and engineer from the local radio station, KNOM. A total of three cars went out to the site and were not disappointed, as the strange luminous being once more made an appearance. When one of the vehicles deliberately tried to run the creature over, the car seemed to pass through it with no effect of any kind.

Most of the same witnesses, along with a large group of other townspeople, returned for yet another night of viewing on August 26, and this time, they sighted three of the little men, whose luminosity varied in color from green to silver to black to blue. The same phenomenon was seen again the following night, on August 27, except only two little men appeared. It was to be the last night they would be seen, although for some days, townspeople gathered at the same location, hoping to catch another glimpse of the luminous little beings.

In researching this story, the authors were hoping to contact reporter Janet Ahmasuk, but she

EARLY AMERICAN UFOS: 1800-1864

unfortunately passed away in 2012. The authors assume that Lois Foster has also passed away, although we were unable to find a record of her death. We established that Lois lived in Pateros, Washington until 1979 and that she often exhibited Eskimo handcrafted items, gave presentations about her Eskimo heritage, and told Eskimo stories to local schoolchildren. In 1979, she and her husband sold their business in order to leave Washington and return to her native village of Koyuk. The local newspaper said, "The Fosters are on their way to Koyuk, Alaska, to start commercial fishing, something they have always wanted to do." Unfortunately, our research turned up no further mention of Lois Foster after 1979.

With the passing of the two persons most directly involved with these amazing "little people" sightings in Alaska, we may never have any additional information about these cases. Still, the historical record seems to strongly suggest that the 1805 Koyuk "little men" incident may have been a true event, based on the testimony of an actual firsthand eyewitness, Lois Foster's grandmother Gadayroak, and vetted by a respected Alaskan journalist, Janet Ahmasuk.

THE REAL COWBOYS & ALIENS

LEWIS AND CLARK AND THE LITTLE MEN FROM MARS?

August 25, 1804
Vermillion River, Dakota Territory

MOST NATIVE AMERICAN tribes have stories about "Sky People" or "Star People" who visited them in the distant past and acted as teachers, guides, and spiritual leaders. So profound is their connection with these beings from the stars that many tribes believe they are the descendants of these Star People. It is often said that to Native Americans, UFOs are nothing surprising, because they have always felt connected to outer space and to the inhabitants of other worlds.

Native American stories also told of "Little People" who imparted spiritual wisdom to the Crow Nation and to other tribes of the West. These little humanoids, known to the Crows as the *Nirumbee* and the *Awwakkulé*, were ferocious

THE REAL COWBOYS & ALIENS

warriors, despite their small stature. These fearsome "spirit dwarves" or "demons" were said to be no more than 18 inches in height and have very large heads in proportion to their bodies. They also gave no quarter to their enemies, slaying them by the hundreds using sophisticated arrows that killed from extremely long range.

In August 1804, these strange humanoids were first brought to the attention of the Lewis and Clark Expedition, the U.S. government's first official effort to explore the western part of the continent, commanded by Captain Meriwether Lewis and his close friend Second Lieutenant William Clark.

The Lewis and Clark Expedition was traveling for a time with a band of Wičhíyena Sioux on the Vermillion River in modern-day South Dakota. On August 25, 1804, Lewis, Clark, and ten other men traveled about nine miles north of the river's junction with the Missouri River to see the "mountain [mound] of the Little People." Lewis wrote in his journal that the Little People were "devils" with very large heads, about 18 inches in height, and very alert to any intrusions into their territory.

The Sioux said that the devils carried sharp arrows which could strike at a very long distance, and that they killed anyone who approached their mound, referred to as the "Spirit Mound." The Little People so terrified the local population, Lewis reported, that the Maha (Omaha), Ottoes (Otoe), and Sioux would not go near the place. The place of the Little People was in the Pryor Mountains, a small mountain range in present-day

EARLY AMERICAN UFOS: 1800-1864

Carbon County, Montana and Bighorn County, Montana.

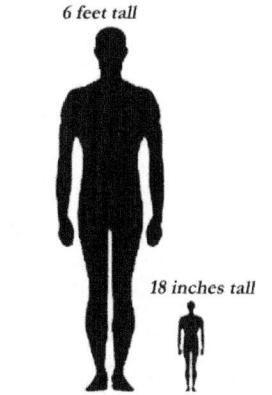

The Little People's Height Compared to a 6' Man

Lieutenant Clark's diary for August 24 reads as follows [corrected for grammar and spelling]: "Captain Lewis and I decided to visit a high hill situated in an immense plain three leagues N. 20° W. from the mouth of White Stone River. This hill appears to be of a conic form, and by all the different nations in this quarter is supposed to be a place of devils, or that they are in human form with remarkable large heads and about 18 inches tall; that they remarkably are very watchful and are armed with sharp arrows with which they can kill at a great distance; they are said to kill all persons who are so hardy as to attempt to approach the hill. They state that tradition informs them that many Indians have suffered by these little people and among others that three Omaha men fell a sacrifice

THE REAL COWBOYS & ALIENS

to their merciless fury not many years since – so much do the Omaha, Sioux, Otoe, and other neighboring nations believe this fable that no consideration is sufficient to induce them to approach this hill."

Chief of the Crow Nation, Plenty Cous

Interestingly, the shortest human to ever exist was 21 inches tall and was an extremely rare example of dwarfism. For an entire group of individuals living in one particular area to all be no taller than 18 inches seems highly beyond the norm.

The chief of the Crow Nation, Plenty Cous (*Aleek-chea-ahoosh*), was said to have several visions and dreams from the Little People, beginning at age nine, that significantly shaped his

EARLY AMERICAN UFOS: 1800-1864

future leadership of the tribe. The spiritual insight received from the Little People led directly to the Crow Nation remaining strong even after most of the other tribes were broken and scattered across North America.

The visions influenced Plenty Cous to promote education for his people and to try to keep alive the culture and beliefs of the Crow well into the future, while other tribes lost so much of their cultural identities. In one of the most influential of these visions, the Chief of the Little People revealed to Plenty Cous that the white men would swarm over the land, but that if he took certain steps, his tribe would survive into the future.

The vision showed that "the day of the Plains Indian was ending, and that white men would swarm over the land like buffalo. But the chickadee [a very small bird representing the Crow Nation] remains, because it is a good listener, develops its mind, and survives by its wits."

It seems that the mysterious Little People had a means of knowing the future, and as a result of their warning, Plenty Coups came to believe that the Crow could survive the coming tide of white people, if his people embraced education and sharpened their mental abilities. He also believed that the Crow would inherit the land in the vicinity of the mound of the Little People.

The Crow Nation *did* survive, and today the Crow Indian Reservation is only a short distance from the Pryor Mountains. A historian has commented, "Indeed, the Crow people survived

THE REAL COWBOYS & ALIENS

the deepest crisis of the nineteenth century in part because of Plenty Coup's vision."

The Pryor Mountains of Montana, Home of the Little People

Not much is known about the culture of the Little People; however, a recurring theme is their great strength, despite their small size. When the Little People felt threatened by the presence of 350 Lakota warriors who approached their territory, the fearsome dwarves almost entirely wiped out the Lakota, leaving behind only a few men who were crippled for life. In another story of their strength, the Crow told of a Little Person who killed a full-grown bull elk and carried it off just by tossing the elk's head over its shoulder.

The Crow created the expression "strong as a dwarf" inspired by the amazing strength of the Little People. These incredibly fierce warriors were said to feed primarily on meat and to have many sharp, canine-like teeth in their mouths. In further evidence of their strength, the Little People were said to tear the still-beating hearts out of their

EARLY AMERICAN UFOS: 1800-1864

enemies' horses. Also, the Crow believed that if anybody offended the Little People, they would utterly destroy both the person who offended them and his entire family.

Members of the Crow Nation (1840-1843)

In addition to their great strength, the Little People practiced "powerful medicine," according to the Crow. In one story, a young Crow boy accidentally fell into a bonfire, leaving his face horribly scarred. He was given the name "Burnt Face" after the accident. Time passed, and the boy decided to approach the place of the Little People, where he was eventually confronted by several of the dwarves. After talking with the boy, the Little People ministered to his scars, causing them to completely disappear. In addition, they gave him healing powers to help his people. When Burnt

THE REAL COWBOYS & ALIENS

Face returned to his people, he retained his name, but his newfound skills enabled him to become a great chief among his people.

In another story, as his family was traveling past the vicinity of the Little People's dwelling place, a small child fell out of the "drag sled" being used to carry him. His family did not notice, and the child was soon totally on his own, but not for long. The Little People took the boy in and taught him their ways. He learned their magic and grew to become very strong in their powers. Using those powers, he stacked up huge rocks into columns and pillars, just for fun, thereby creating the geologic formation known today as Medicine Rocks.

Yet another story of the Little People has a hunter going on a hunting expedition in the Pryor Mountains but has little luck. He asks the Little People for guidance and is told that he must first provide an offering. After shooting a deer and presenting it as an offering, his luck at hunting totally changes. Curious to see what happened to the deer he offered up to the Little People, he returned to the spot of the offering, only to find the deer's corpse vanished.

People wonder if archeologists have ever found evidence of the culture of the Little People in the area of the Pryor Mountains, and the answer is yes. The physical remains of very small people were reportedly found in various locations in Montana. These remains were found in caves and were described as being "perfectly formed," dwarf-size humans. Unfortunately, the specimens collected were lost by researchers before a complete analysis

EARLY AMERICAN UFOS: 1800-1864

could be done. Some skeptics suggest that the "tiny humans" could have actually been anencephalic infants or infants with other deformities.

As we close our look at the strange Little People of the Pryor Mountains, it is interesting to review the observations about them. These amazing humanoids reportedly had: (1) a stature of about 18 inches, (2) abnormally large heads, (3) healing powers, (4) prophetic powers, (5) weapons of advanced technology, (6) the ability to communicate with humans through visions and dreams, (7) physical and mental strength such that all the tribes around them avoided them, and (8) an unknown origin and final disposition.

All of this seems to be compelling evidence that these tiny humans were separate and apart from the rest of humanity. Might the Little People have been a crew of extraterrestrials that had been stranded on Earth? Might they have been exiled here? If not extraterrestrials, perhaps they were members of an unknown civilization that remains hidden to us, even today?

THE REAL COWBOYS & ALIENS

Sophie Anderson's 1869 Painting of a Fairy

4
UFOS CROP CIRCLES AND FAIRIES

July 1806
Providence, Rhode Island

THOUGH FAIRIES may not sound intimidating thanks to their popularization in literature and film (like Tinker Bell in *Peter Pan*), fairies have a dark side. The popular television series *Supernatural* included a sixth-season episode called "Clap Your Hands If You Believe," which links UFOs to these "dark fairies." While investigating a rash of alien abduction cases in Elwood, Indiana, Dean Winchester finds himself "abducted" in a crop circle, but not by what we traditionally call "aliens." The aliens are in fact fairies that disguise themselves as aliens in order to correspond with modern beliefs.

Encounters with fairies are strikingly similar to accounts of alien abductions. In many of these

THE REAL COWBOYS & ALIENS

narratives, the witness encounters a glowing orb of light. They are then often rendered unconscious, waking up hours later with no memory of what happened to them -- what we would today call "missing time." In cases where the victims do remember what happened, often their encounters with the "little people" or fairy folk include an "examination of their bodies" that seemed sexual in nature, similar to what is often reported during alien abductions.

Consider this example: A man driving his car along the highway sees a mysterious, bright light. He wakes up hours later, with no clear memory of what has occurred. Days, months or years later he has vague memories of being examined by strange beings. Just swap out the car for a horse and buggy, and you have a fairy encounter of the old days.

And then there are "fairy rings," a literal ring of grass or mushrooms which have grown in the shape of a perfect circle. Ancient Europeans, and even more recent believers, considered them to be spots of supernatural significance that could lead to good or bad [but mostly bad] luck. In some cases, people claimed to see fairies dancing within the fairy rings. In some ways, these fairy rings are eerily similar to the UFO-related "crop circles" of today.

Take for example this old Scottish poem espousing the dangers of fairy rings:

He who plows the fairies' ground
Never again shall have good luck
And he who destroys the fairies' ring
Brings upon himself need and sorrow,

EARLY AMERICAN UFOS: 1800-1864

*For days without magic and weary nights
Are his until his dying day.*

*But he who passes by the fairy ring,
Neither suffering nor longing shall see,
And he who cleans the fairy ring,
An easy death shall die.*

*17th Century Engraving,
of a Fairy Ring and Fairies Within*

One of the more disturbing "fairy tales" -- and we mean that literally, as in a story involving a potentially real fairy --comes from Wales. It concerns a beautiful teenage girl named Shui Rlys, a farm girl with no interest in farm chores. Instead, she would disappear each day for hours on end only to return home to her angry mother, who naturally scolded her. Eventually Shui revealed that

THE REAL COWBOYS & ALIENS

she was with the *Tylwyth Teg* (Welsh forest fairies dressed in green clothing), who kept her spellbound with strange music from their harps. For fear of the magical fairies, Shui's mother no longer scolded her. Then one day Shui never came home again, the belief being that the fairies had carried her away.

A Fairy Ring (from the 1880 book "British Goblins: Welsh Folk-lore, Fairy Mythology, Legends and Traditions")

The really interesting detail here is the green clothing. Interestingly, the term "Little Green Men" did not originate with aliens; it originated with fairies. Many fairies were said to be clothed in green, and not just the Leprechauns. For instance, something sounding very much like a grey alien, either wearing all green or green skinned, was killed in Texas in 1913 (a story for another book, covering another time period).

Many of these little people were in fact described as having "big eyes" -- but unlike the grey aliens

EARLY AMERICAN UFOS: 1800-1864

seen since the 1940s, many of them also had facial hair. As just one example of this, there is the account of Bob McCain a fur trapper from New York State in the 1830s. McCain claimed that he had rescued a little man in an all green suit with big eyes and no eyebrows from a trap near a river. The little man was said to be weightless, and he ran away from McCain upon being freed.

There is also an 1853 alien-like fairy encounter from Washington Island, Wisconsin. An eight-year-old girl was out in the fields collecting wild berries when she was "accosted" by fairies. In this case the fairies were described as being tiny human-like creatures that wore "shiny clothing" with pointy hats, much like a classic lawn gnome. They begged the girl for some berries then allowed her to dance with them. They also taught her a new song with a "haunting melody." In light of the parallels between aliens and fairies, we could then ask, were the "little people" in these stories spectral or extraterrestrial? Some researchers have suggested that the beings we now call grey aliens in previous centuries appeared as fairies and other "mythical" creatures in order to "fit" into the legends and fables of the time. What is important, they say, is not their appearance, but their actions. Whether fairies or aliens, they abducted people, conducted examinations of humans, appeared in conjunction with "orbs" and other anomalous objects, and exhibited a high level of intelligence or sophistication, interpreted as "magic" by the simple people of the time.

THE REAL COWBOYS & ALIENS

Stories about fairies or strange "little people" were not unique to Western Europe; they also took place in North America. Hopkinton, New Hampshire was in particular steeped in lore regarding ghosts and witches. One man, in all earnestness, believed that a witch riding her broom had knocked some shingles off his roof. The book *Life and Times in Hopkinton, N.H.* by Charles Chase Lord tells of a few "fairy" encounters from the 1820s that sound like they could actually be UFO sightings.

Lord writes, "Upon the northern brow of Putney's hill, sometimes known as Gould's hill, is a patch of forest long recognized as the 'Lookout,' once the point from which observations for possible locations of Indians was taken, the smoke of fires revealing their haunts. Spectral appearances in different forms, manifested both by day and by night, were apprehended in this locality. The writer remembers a respectable man who believed to his dying day that he there saw an apparition in broad daylight. There was living in this town recently an old and respectable gentleman who once averred that passing the Lookout in the evening, returning from his day's work, he saw several balls of spectral fire appear and stand before him, keeping in his advance as he maintained his distressful march home."

These balls of fire, one could argue, sound either like small UFOs or fairies. It seems as if the narrator of the above story narrowly avoided some sort of abduction, be it by fairy or alien - if the two are not in fact the same. Renowned UFO

EARLY AMERICAN UFOS: 1800-1864

researcher Jacques Vallee noted in his book, *Wonders in the Sky: Unidentified Aerial Objects from Antiquity to Modern Times*, "Such glowing balls had been seen in the area since 1750, moving slowly in mid-air, so one may suspect a natural phenomenon."

An even more fairy-like *Close Encounter of the Third Kind*, or UFO encounter involving the sighting of animated beings, took place in Providence, Rhode Island in 1806. Though this is thought of as a ghost story by many, we shall view it through the lens of both Ufology and the paranormal in regard to fairy lore.

The story was recorded in the pamphlet "Immortality proved by Testimony of Sense" by the witness himself, the Reverend Abraham Cummings. The Baptist preacher came to the fishing village via sailboat. Described as an "otherworldly" and "absent minded" individual, Cummings had a keen interest in the spirit realm. Since 1799, the area had been subject to a great deal of paranormal activity, which included many people hearing an unseen woman's voice. The identity of the woman could never be pinned down to one individual, but many thought she was the dead wife of George Butler. This explanation did not convince the dead woman's own sister who -- even upon hearing the ghost's voice -- believed another entity was merely impersonating her and that it was not her dear sister's spirit, but an impostor.

THE REAL COWBOYS & ALIENS

Reverend Cummings saw the spectral woman in the summer of 1806, on an unspecified date during the month of July. Though he was a skeptic, when the woman was sighted at the Blaisdel residence he decided to go see for himself. Upon walking out their front door to the area of the recent sighting, he spied some white rocks strewn atop a knoll sticking out of the ground. He assumed that these ghostly white rocks may have been what confused the simple village people.

Then, one of the rocks began to levitate and turned into a glowing ball of light -- much like a fairy. The white globe flashed red and then took on the form and proportions of a fully grown woman. However, this figure had the height of what Cummings compared to a "seven-year-old girl." [A floating grown woman of small stature -- that certainly fits the profile of a fairy.] Cummings thought to himself, "You are not tall enough for the woman who has been appearing among us." At that point the figure expanded to the size of a normal grown woman. In his pamphlet, Cummings wrote that she appeared "glorious," and that rays of light shone from her head. The reverend recorded that, although terrified, "my fear was connected with ineffable pleasure." This too is similar to fairy accounts, and also certain alien abduction accounts.

Cummings wrote that after that miraculous vision, everything else in his life seemed mundane by comparison. For the month of August, the fairy woman made regular appearances, seeming to give the impression of benevolence. Tales of the

EARLY AMERICAN UFOS: 1800-1864

sightings became so widespread that soon visitors would pack into the Blaisdel family cellar to meet the apparition.

The specter's activity was as much in line with that of a fairy as it was that of a ghost. If grey aliens can transform themselves either by their mental abilities or other forms of disguise, this could easily be interpreted as a UFO incident. Even in modern times, aliens do not always appear as "greys"; the second most widely reported alien form are the human-like "Nordics," which are generally tall, blonde and often attractive. Considering this figure was said to be literally radiant in her beauty, is it so far off to speculate that the entity could have been an alien in Nordic form? Furthermore, though initially frightened, many alien abductees consider the beings to be friendly and benevolent, much like fairies.

In any event, the connection between old fairy folklore and alien abductions certainly deserves additional research from the field of Ufology. Researchers continue to express fascination at how ancient stories about fairies seem to so closely resemble modern eyewitness testimony about UFOs and their strange occupants.

THE REAL COWBOYS & ALIENS

For the Raleigh Register.

EXTRAORDINARY PHENOMENON.

The following account of an extraordinary phenomenon, that appeared to a number of people in the County of Rutherford, State of North-Carolina, was made the 7th of August 1806, in presence of David Dickie, Esq. of the county and State aforesaid, Jesse Anderson and the Rev. George Newton, of the County of Buncombe, and Miss Betsey Newton of the State of Georgia, who unanimously agreed, with the consent of the relaters, that Mr. Newton should communicate it to Mr. J. Gales, Editor of the Raleigh Register and State Gazette.

Patsey Reaves, a widow woman, who lives near the Apalachian Mountain, declared, that on the 31st day of July last, about 6 o'clock P. M. her daughter Elizabeth, about eight years old, was in the Cottonfield, about 10 poles from the dwelling house, which stands by computation, 6 furlongs from the Chimney Mountain, and that Elizabeth told her brother Morgan, aged 11 years, that there was a man on the mountain. Morgan was incredulous at first; but the little girl affirmed it, and said she saw him rolling rocks or picking up sticks, adding that she saw a heap of people. Morgan then went to the place where she was, and calling out, said that he saw a thousand or ten thousand things flying in the air. On which, Polly, daughter of Mrs. Reaves, aged 14 years, and a negro woman, ranto the children, and called to Mrs. Reaves to come and see what a sight yonder was. Mrs. Reaves says, she went about 8 poles towards them, and, without any sensible alarm or fright, she turned towards the Chimney Mountain, & discovered a very numerous croud of beings resembling the human species; but could not discern any particular members of the human body, nor distinction of sexes; that they were of every size, from the tallest men down to the least infants, that there were more of the small than of the full grown, that they were all clad with brilliant white raiment, but could not describe any form of their raiment; that they appeared to rise off the side of a mountain south of said rock, and about as high; that a considerable part of the mountain's top was visible above this shining host, that they moved in a northern direction, and collected about the top of the Chimney rock. When all but a few had reached said rock, two seemed to rise together, and behind them about two feet, a third rose. These three moved with great agility towards the croud, and had the nearest resemblance to men of any before seen. While beholding those three, her eyes were attracted by three more rising nearly from the same place, and moving swiftly in the same order and direction. After these, several others rose and went towards the rock.

During this view, which all the spectators thought lasted upwards of an hour, she sent for Mr. Robert Siercy, who did not come at first; on a second message sent about fifteen minutes after the first, Mr. Siercy came; and being now before us, he gives the following relation, to the substance of which Mrs. Reaves agrees.

Portion of Original Article from the Weekly Raleigh (N.C.) Register, September 15, 1806

THE FLYING HUMANOIDS
July 31, 1806
Chimney Rock, North Carolina

EARLY IN THE 21st century, many UFO reports surfaced in the U.S. and Mexico involving not a craft, such as a metallic disc or sphere, but rather simply one or more humanoids flying through the air using an unknown means of propulsion. The cryptid website *Exemplore.com* notes, "Sightings of flying humanoids are truly one of our world's most bizarre unsolved mysteries. The narrative is usually as simple as it sounds. Witnesses report strange objects in the sky, but not the usual UFOs or flying saucers. They're seeing humans, or human-like creatures, suspended in air, apparently flying on their own."

In the 1800s, many similar "flying humanoid" reports were filed from throughout North

THE REAL COWBOYS & ALIENS

America, including a very unusual case that occurred in the Appalachian Mountain region of North Carolina on July 31, 1806. Five witnesses saw "throngs" of flying humanoids wearing "brilliant white raiment" hovering on Chimney Mountain (now called Chimney Rock, elevation 2,280 feet). The strange sighting, which lasted over an hour, was reported to the editor of the *Weekly Raleigh Register* in Raleigh, North Carolina, and was published in the paper on September 15, 1806. The account was also printed in several other U.S. newspapers as far away as Pittsburgh, Pennsylvania. The principal witness in the case was a widow named Patsey Reeves, living in a house located approximately three-quarters of a mile from Chimney Rock. Also living with her were her children: Polly, 14; Morgan, 11; Elizabeth, 8; and an African American servant whose name is not given in the story. A fifth witness, Robert Siercy (or Searcy), joined the other witnesses near the end of the sighting.

At about 6 p.m. on July 31, 1806, Morgan and Elizabeth were standing in a cotton field located about 50 yards from the house, when they saw a very unusual sight on the nearby rock. The two children observed a group of strange humanoids, described as "a heap of people," hovering near the top of the rock, seemingly gathering vegetation and moving rocks around. Morgan later stated that, in addition to the creatures that were "rolling rocks or picking up sticks," a multitude of others were "flying in the air" near the top of the rock. The children were soon joined by Polly and the servant,

EARLY AMERICAN UFOS: 1800-1864

who after confirming the spectacle, called out to Mrs. Reeves to witness the sight. From about 40 yards away, Mrs. Reeves turned toward Chimney Rock and saw the swarm of flying humanoids with her own eyes. She stated that the creatures were "clad with brilliant white raiment," but could not describe any specific details about their clothing.

Recent Photo of Chimney Rock by Ehume
[CC BY-SA 4.0 (https://creativecommons.org/licenses/by-sa/4.0)]

According to Mrs. Reeves, the strange creatures resembled "the human species," but she "could not discern any particular members [defining characteristics] of the human body, nor distinction of sexes." She also stated that "they were of every size, from the tallest men down to the least infants" but "there were more of the small than of the full grown." If filtered through the lens of Christianity, this might be viewed as a heavenly apparition of a company of saints, but unfiltered, it seems there was a massive search underway near the top of Chimney Rock by a large group of unknown

THE REAL COWBOYS & ALIENS

creatures using an advanced personal propulsion system to move through the air with great ease.

For about an hour, the assembled witnesses watched as the creatures moved "swiftly" up and down Chimney Rock "with great agility," typically in groups of two or three. The creatures' movement was generally toward the top of the rock, although at one point, three of the beings "moved with great agility towards the crowd," which allowed the spectators the best view of them as any during the entire sighting. Perhaps these three were sent to assess whether the spectators might be a threat to the search party.

After having watched the strange spectacle for some time, Mrs. Reaves sent someone from her group to summon a neighbor, identified as Mr. Robert Siercy (most likely a misspelling of Searcy). Upon arriving at the scene, a short time later, Siercy said he saw "more glittering white appearances of humankind than ever he had seen..." He confirmed that the strange visitors were of different sizes, ranging from men to infants. He stated that in their hovering around Chimney Rock, they never flew higher than the height of the mountain. It was Siercy who gave the only account of how the sighting ended, noting that two of the creatures moved toward the crowd of spectators, coming within 20 yards of them, when suddenly, they "vanished out of sight, leaving a solemn and pleasing impression on the mind. Accompanied with a diminution of bodily strength." The actual manner of their vanishing is not described, and it is

EARLY AMERICAN UFOS: 1800-1864

possible that the strangers just rapidly flew away from the crowd until they were all "out of sight."

The story, based on the direct testimony of the eyewitnesses, was assembled and submitted to the newspaper by George Newton (1765-1840) a Presbyterian minister assigned to several small congregations in Buncombe County, N. Carolina.[5] This would explain the "ecclesiastical" language that appears in the story such as "brilliant white raiment," which could also describe clothing typically worn by the angels or saints. He also used the phrase "shining host" as in describing a "host of angels." And, he closes the narrative by writing, "... whether it be a prelude to the descent of the Holy City, I leave to the impartially curious to judge." When analyzing that facts of the story, one must take into account Newton's attempt to frame the encounter in the language of Christianity. If one corrects for this tendency, the narrative becomes quite interesting in its secular details. The amazing technology exhibited by the flying humanoids would certainly appear to be magical or mystical to the people of 1806, especially to a minister of the Presbyterian Church. As Sir Arthur C. Clarke put it many years later, "Any sufficiently advanced technology is indistinguishable from magic."

The article states that the eyewitnesses were urged to tell their story to the world by Reverend Newton; Betsey Newton of Georgia; David Dickie, Esq.; and Jesse Anderson. All were unanimously agreed that the story should be published, and they

---

[5] https://www.ncpedia.org

THE REAL COWBOYS & ALIENS

received permission from Patsey Reeves to communicate it to Mr. J. Gales, editor of the Raleigh Register and State Gazette. The editor's identity has been confirmed by viewing microfilm back issues of the newspaper from 1806. Also, the identities of most of the named persons have been confirmed by historical records, including witness Robert Siercy (Searcy), who lived near Chimney Rock in 1806, according to the Searcy family ancestry web page.[6]

Were the Chimney Rock Humanoids Wearing Something Like Modern Hazmat Suits? (Public Domain)

---

[6] http://www.searcyfea.com

EARLY AMERICAN UFOS: 1800-1864

Looking at this case more than 200 years after it happened and through the lens of a 21st century understanding of technology, the "brilliant white raiment" worn by the strange creatures was certainly some kind of suit, such as a space suit, hazmat suit, or biohazard suit. These types of suits would make it hard for the witnesses to distinguish very much detail about the bodies of the humanoids, especially their gender. The suits were also obviously equipped with some type of propulsive backpack that allowed the beings to hover above the terrain at will. It seems clear that the strangers were conducting a search for something in and around Chimney Rock. They finally departed either because they found what they were looking for, gave up the search, or became concerned at the growing number of human spectators.

As to what was the object of their search, one can only speculate – perhaps certain minerals or elements found in the area? Scientists say the Chimney Rock formations are more than 500 million years old and are composed mostly of granitic gneiss, containing deposits of feldspar, mica, and quartz. The rocks are also known to contain trace amounts of radioactive materials, which can sometimes cause elevated levels of radon in local groundwater.

Finally, in closing this interesting tale, we should note that over the years since 1806, the area around Chimney Rock has seen frequent UFO sightings. Reported to the Mutual UFO Network (MUFON) on May 8, 2012 was this incident at Chimney Rock:

THE REAL COWBOYS & ALIENS

"Fiancé and I were driving down a dirt road and noticed what looked like a large planet-sized light. It split into 3 separate lights; one headed right, one hovered in place, and another to the left. The left one separated for about 20 seconds then merged with the middle light, and the right light went in a straight line toward the trees. They vanished as if all lights simultaneously shut off at the same time." Another incident, on July 11, 2015, is described as follows: "We owned an ice cream shop with outside seating in downtown Chimney Rock, NC. Around 10 p.m. we decided to stay open late since one couple was dancing to the beach music we were playing, and two other couples were sitting enjoying the evening. My husband and I joined our customers and were sitting down when a bright yellow/white light darted from the range of mountains across the street and was chased by a military helicopter. The chase lasted only a few seconds as the UFO shot across the sky toward the lighted Chimney Rock (a chimney shaped rock) that was on the opposite side of the road (our side). The two objects soared around the 'Chimney' and the UFO made a hard-right turn with the helicopter on its tail in pursuit. It was unmistakably a UFO with a military helicopter giving a hard chase. Both disappeared over the range of mountains that comprise the park. It was a clear night, stars were out, no moon. The only sound was that of the helicopter. The helicopter was right on the tail of the UFO by the time the two reached the 'Chimney.' It was an incredible sight."

EARLY AMERICAN UFOS: 1800-1864

Perhaps the ETs are still searching for something at Chimney Rock, even after all these years!

Were the Humanoids Using Some Type of Jet Pack, such as this One Tested by NASA and USGS, circa 1966?

THE REAL COWBOYS & ALIENS

UFO FROM THE PAST?–Dr. Judith Becker Ranlett examines the diary written by her husband's great-great grandmother, which contains a passage describing a strange light that appeared in the sky in Camden, Maine, in 1808. The light moved erratically. Dr. Ranlett, who is studying the diary, thinks the passage described a UFO. (Hal Stokes photo).

Diary Describes UFO Seen In 1808

BY HAL STOKES

Back in 1808 in Camden, Maine, there certainly were no weather balloons or Air Force jets to be confused with flying saucers. George Washington's soldiers had barely gotten back home to their farms.

But something odd happened one summer night that year which was recorded in the diary of a Potsdam man's great-great grandmother.

Today the passage is interpreted by his wife, an historian who is studying the diary, as a first-hand account of a UFO sighting in the early 19th century.

"I thought she was describing a UFO when I first read it," said Dr. Judith Becker Ranlett, an historian who teaches at the State University College at Potsdam. "If she had seen something normal, she would have attempted to explain it as a natural phenomenon," according to Dr. Ranlett, who is using appearance. It was a light which proceeded from the East. At the first sight, I thought it was a Metier, but from its motion I soon perceived it was not. I(t) seem to dart at first as quickly as light; and appeared to be in the Atmosphere, but lowered toward the ground and kept on at an equal distance sometimes ascending and sometimes descending. It moved round in the then visable Horison, (it was not very light) and then returned back again, nor did we view it till it was extinguished."

That is the only passage in the entire diary that mentions the sighting, according to Dr. Ranlett. She finds it significant that Cynthia Everett did not explain what she witnessed as a natural phenomenon, since she was well-educated and had first-hand knowledge about the night sky. "She was the kind of person who would have explained it as a natural phenomenon, if she could have."

Dr. Ranlett reasoned, because Cynthia would have been teaching school at 10 a.m. and besides she always made her entries in the diary just before she went to bed. Dr. Ranlett said she determined that the sighting was in Camden through the various people that are referred to on that day.

Cynthia was 24 years old when she wrote about seeing the strange light. She was single but was living, as teachers did, with a family in the area of the school. She changed her lodgings about once a week, according to Dr. Ranlett.

The schoolteacher had a good education for the period, Dr. Ranlett said. She had attended Leicester Academy at Leicester, Mass., one of the few truly coeducational schools where women went to class with men.

The diary was written until Cynthia was in her 30th year. Entries cease three days after her marriage to John Ranlett,

Portion of the Original Newspaper Article in the Potsdam, N.Y. Courier & Freeman, March 28, 1978

6

THE SCHOOL TEACHER AND THE UFO

July 22, 1808
Camden, Maine

RESEARCHERS of old UFO cases dream of opening the diary or journal of someone who lived hundreds of years ago and suddenly discovering that they wrote down a first-hand, eyewitness account of a UFO encounter that occurred in the distant past. Although she is not a UFO researcher, this is exactly what happened to historian Dr. Judith Becker Ranlett in 1978, while reading through the 600 handwritten pages that comprised the diary of her husband's great-grandmother, Cynthia Everett. Although the purpose of her research was not UFOs and she had no real interest in the topic, she realized that what she found in the diary was an important record of a significant, unexplained early UFO sighting. Consequently,

THE REAL COWBOYS & ALIENS

she decided to publish the section of the old diary concerning the incident, shown below with grammatical and spelling errors corrected:

Friday, July 22, 1808 -- "About 10 o'clock I saw a very strange appearance. It was a light which proceeded from the east. At the first sight, I thought it was a meteor, but from its motion I soon perceived it was not. It seemed to dart at first as quickly as light. And appeared to be in the atmosphere but lowered toward the ground and kept on at an equal distance sometimes ascending and sometimes descending. It moved round in the then visible horizon, it was not very light, and then returned back again, nor did we view it till it was extinguished."

In carefully analyzing Everett's short diary entry and breaking it down into its component phrases, we discover some important details about what she saw: "About 10 o'clock I saw a very strange appearance. It was a light which proceeded from the east." From this part of the entry, we discover that the sighting occurred at 10 p.m., which on July 22, 1808 would have been 3 ½ hours after sunset (7:24 p.m.). We also learn that the glowing object was inbound to Camden from the east, which means that it came across the North Atlantic Ocean, possibly over Sable Island and across the southernmost land mass of Nova Scotia, before appearing in Camden. Dr. Ranlett's comment: "At the first sight, I thought it was a meteor, but from its motion I soon perceived it was not." Everett's knowledge of science and the world around her kicked in, and she immediately began trying to

EARLY AMERICAN UFOS: 1800-1864

scientifically analyze what she was looking at. Everett had a good education for a woman of that time period, Dr. Ranlett said. She had attended Leicester Academy at Leicester, Massachusetts, one of America's earliest co-educational schools. Her first thought was that she was looking at a meteor, but she quickly dismissed it because of the erratic motion of the object.

Typical 19th Century Maine Schoolhouse
(BMRR from en.wikipedia [CC BY-SA 3.0 (https://creativecommons.org/licenses/by-sa/3.0)]

At the time the diary entry was written, Cynthia Everett was a 24-year-old schoolteacher in Maine, and Dr. Ranlett says that she was well familiar with scientific topics, including the night sky, earthquakes, comets, etc. Elsewhere in her diary, Everett records the occurrence of an earthquake and the passing of a comet. But in the case of what she saw on July 22, 1808, she drew a complete blank on what the object might have been.

THE REAL COWBOYS & ALIENS

Dr. Ranlett said, "I thought she was describing a UFO when I first read it. If she had seen something normal, she would have attempted to explain it as a natural phenomenon. She was the kind of person who would have explained it as a natural phenomenon, if she could have. In fact, she did, her first thought was that it was a meteor."

"It seemed to dart at first as quickly as light. And appeared to be in the atmosphere but lowered toward the ground and kept on at an equal distance sometimes ascending and sometimes descending." The motion of the object is the key to understanding how unusual this sighting was. Everett observed that the glowing light remained in a stationary position relative to the ground, but it moved up and down. At first it was at a very high altitude ("in the atmosphere") and later approached close to the ground.

"It moved round in the then visible horizon (it was not very light) and then returned back again, nor did we view it till it was extinguished." From Everett's description, it appears that the light danced along the horizon for some time, moving up and down, as well as in a circle. The comment "nor did we view it till it was extinguished" indicates that there were other observers besides Everett and that, at some point, they called it a night and stopped watching the spectacle.

In an article that appeared in the Potsdam, N.Y. *Courier & Freeman* newspaper on March 28, 1978, reporter Hal Stokes wrote, "The sighting must have been at night, Dr. Ranlett reasoned, because Cynthia would have been teaching school

EARLY AMERICAN UFOS: 1800-1864

at 10 a.m. and besides she always made her entries in the diary just before she went to bed. Dr. Ranlett said she determined that the sighting was in Camden through the various people that are referred to on that day. Cynthia was 24 years old when she wrote about seeing the strange light. She was single but was living, as teachers did, with a family in the area of the school. She changed her lodgings about once a week, according to Dr. Ranlett."

In Cynthia Everett's very short diary entry is an amazing early document of something highly unusual, and it happened in an area of the United States that has been a hotbed of UFO activity over the decades since 1808. In 2019, the web site *SatelliteInternet.com* found that Maine has among the highest number of reported UFO sightings per capita in the U.S. According to statistics from the National UFO Reporting Center and the U.S. Census Bureau, Maine experiences 70.23 sightings per 100,000 people. In fact, over the years, Maine has consistently ranked in the top five for per capita UFO sightings. A reporter for the Portland, Maine *Press Herald* newspaper recently wrote, "Maine is known for many things -- lobster, blueberries, scenic vacations. But UFO sightings? According to a compilation of reported sightings of unidentified flying objects from around the country, Maine hosts among the highest [UFO sightings per capita.]"

Although it did not involve a UFO sighting, a very strange unexplained event happened a few years later just nineteen miles away from Camden in the nearby community of Waldoboro, Maine. As will be discussed in an upcoming chapter, a small humanoid was discovered and captured in the woods of Waldoboro, causing a nationwide media sensation. The story of the "Waldoboro Wild Man" appeared in newspapers all over the land in 1855.

Betty & Barney Hill (Courtesy Kathleen Marden)

Maine and neighboring areas have also seen some of the most notorious UFO encounters in history, including the 1961 abduction of Betty and Barney Hill in the White Mountains, about 150 miles from Everett's 1808 sighting. While driving in the mountains on the night of September 19, the Hills saw a bright light in the sky that seemed to be pursuing them. At one point, Barney stopped the car and, looking through binoculars, saw an odd-shaped, 40-foot-long UFO with flashing,

EARLY AMERICAN UFOS: 1800-1864

multicolored lights. At a later point, he saw humanoid figures in black uniforms, visible through the windows of the craft. Later, they both "blacked out" and experienced "missing time." It was only later, while under hypnosis, that they both described being taken aboard the UFO, where a series of apparent medical examinations were conducted on them by a group of strange creatures. Their story was adapted by journalist John G. Fuller into the best-selling 1966 book *The Interrupted Journey* and the 1975 television movie *The UFO Incident.*

Also, within 150 miles of Camden is the site of the world-famous Exeter, New Hampshire UFO Incident, which took place on September 3, 1965. The case became the basis for a 1966 bestselling book called *Incident at Exeter,* written by John G. Fuller. The incident began just after midnight when a policeman stopped to check on a woman parked alongside a road. Out of breath and obviously terrified, the woman claimed that a flying object with red flashing lights had been chasing her. A few hours later, an 18-year-old man arrived at the Exeter police station and claimed that while hitchhiking, he had seen a line of five bright lights over a house about 100 feet from where he stood. After driving to the site with the young man, the policeman also witnessed the lights, as did another officer who arrived a short time later. Over the weeks that followed, authorities received about 60 reports of UFO sightings near Exeter.

A bit farther away from Camden, on the northern end of the state of Maine, the infamous "Allagash

THE REAL COWBOYS & ALIENS

Abductions" occurred in August 1976. Four Massachusetts college students were canoeing on Maine's Allagash Wilderness Waterway when they observed an unidentified object in the sky. When they tried to report their sighting to authorities, they were not taken seriously, and they decided not to talk about it anymore. Years later, one of them, Jim Weiner, started experiencing seizures and claimed to see humanoid beings hovering above his bed and poking him with needles at night. Under hypnosis, all four men described small gray aliens taking them aboard a spacecraft and performing medical examinations on them. Later, one of the four men recanted, saying that although they did have a UFO sighting, the abduction claim was not true; however, his three companions still claim the abduction was real.

These are just a few of the most notable UFO cases that have occurred in Maine and surrounding areas since 1806. The sighting by Cynthia Everett was just the beginning of Maine's long history of strange encounters with UFOs and their unearthly occupants. In a way, it was like the opening volley of a mystery that has persisted for over 200 years. Little did Everett realize as she looked at the strange object coming toward Camden from the Atlantic Ocean, that before her eyes was a harbinger of many years of UFO visitations to Maine and to New England in the years and decades to follow.

FLOATING UFO OF THE ATLANTIC

April 8, 1813, Atlantic Ocean

BEFORE EMBARKING on this chapter, it is important to denote that in addition to unidentified flying objects (UFOs), there are also unidentified submersible objects (USOs), which are objects that are observed in or under the water. Though not as highly publicized as their aerial counterparts, USO sightings are more common than one thinks. The naturalist/cryptozoologist Ivan T. Sanderson even wrote a whole book on the phenomena entitled *Invisible Residents: The Reality of Underwater UFOs*. Sanderson rather intelligently connected the dots, stating that if a craft were capable of deep space travel, it would most likely also be capable of travel deep in the ocean. And, what better place for a UFO to hide than under the water?

THE REAL COWBOYS & ALIENS

In this chapter, we will examine perhaps the first USO sighting ever, which occurred in 1813. Since then, there have been many cases of strange objects darting in and out of the water in addition to strange lights seen under the water. A classic example of a more recent USO sighting took place on July 20, 1967, about 120 miles off the coast of Brazil. The sighting was officially included in the log of the ship *Naviero*, which Sanderson quotes in his book:

"Arriving at once on deck, Captain Ardanza beheld a shining object in the sea no more than about 50 feet away on the starboard side. It was cigar shaped and he estimated its length at about 105 to 110 feet. It had a powerful blue and white glow, made no noise whatsoever, and left no wake in the water. There was no sign of any periscope, or railing, or tower, or superstructure; in other words, no external control surfaces or protruding parts."

As the story continues, the mystery craft suddenly dives beneath the ship and vanishes rapidly into the depths at a great speed. All the while the crew could observe its bright glow disappearing under the water. Upon returning to the mainland the crew was absolutely certain that what they had seen was some type of UFO and not any type of strange whale or submarine.

With that example of a typical USO established, we shall now take a look at what could very well be the first recorded USO. The story was printed out of New York under the title of "The Sea Mammoth" in the June 23, 1813 edition of the *Natchez Gazette*. It begins with the following:

EARLY AMERICAN UFOS: 1800-1864

"The subject stated in the subsequent affidavit, having been doubted by many on its first publication, it was thought advisable to bring it forward as it now is, authenticated under the oaths of the three respectable gentlemen whose signatures are affixed to it."

"On 28 April, 1813, before me, the undersigned notary public, personally came and appeared, Samuel G. Bailey, late master of the ship *Amsterdam Packet*, Wm. R. Handy, late master of the ship *Lydia*, and Adam Knox, late master of the schooner *Augusta*, all belonging to New York, and the said deponents being duly sworn according to law, severally and solemnly declared that they were passengers on board the ship *Niagara*, which arrived at this port from Lisbon on Saturday last..."

The article continues that on April 8th, at 43 degrees latitude, 65 degrees longitude at the Meridian, the men sighted "a large lump on the horizon" eight miles away to the northwest. It resembled a great ship turned upside down, or as they put it, "the whole of a large ship bottom upwards." As they got nearer, they were shocked to see that the ship, monster -- or whatever it was -- was capable of movement. What kind of movement, how fast, etc., unfortunately they do not specify. As they got closer, the men began to think it was a massive fish or sea monster which they estimated to be 200 feet in length and 30 feet wide. The parts sticking above the water they estimated to be 17 to 18 feet high in the center. The thing was covered with a shell "formed similar to the planks on a clinker-built vessel." Clinker built refers to a

THE REAL COWBOYS & ALIENS

method of ship building where the edges of the hull planks overlap each other as opposed to being overly smooth.

Underwater Craft Envisioned by Jules Verne, 1896

The description unfortunately becomes a bit puzzling as they next describe something near "the head," without explaining what the head looks like. Presumably, they mean the front of this strange vessel which they supposed to be some sort of mammoth fish, for if it had a monstrous head to observe surely they would have described it. Anyhow, the article continues, "Near the head, on the right side was a large hole or archway, covered occasionally with the fin, which was at times about eight or 10 feet out of the water."

The words "hole" and "archway" would seem to be a dead giveaway that this is a constructed craft, and not a flesh and blood animal even if they do mention a fin. This "fin" could have been a fin for

EARLY AMERICAN UFOS: 1800-1864

the purpose of sailing through the water, or possibly even the air if indeed it was some sort of advanced vessel. Unfortunately, the witnesses don't really give a clear description of the object's shape, but considering they note it as 200 feet long but only 30 feet wide it implies an oblong shape.

Sea Monsters were a Recurring Theme Among Sailors of the 1800s

The article concludes that there was an intention of sending the boat closer for a better inspection, but they were "deterred from the dreadful appearance of the monster, having approached within 30 yards of it."

Usually when reading accounts like this, the obvious question is, "Did that really happen?" In this case, we follow that question with another: "Was it a sea monster?" As for question one, the

THE REAL COWBOYS & ALIENS

fact that this story bears a sworn affidavit means something. But then again, sailormen have always been known for telling "fish stories," as the cliché suggests. But, with that mindset, if you were going to create a tall tale, why fabricate one as weird as this, with no real beginning or end? If making up a story, then why not have the craft chase the boat? If it was a made-up monster, why not describe the head and other body parts as opposed to the curious, vague description we received? Instead, this would seem to bear resemblance to a true sighting, considering all the specific, factual details given, exact latitudes/longitudes, measurements, etc. That all said, unfortunately, we the authors could turn up no further historical records mentioning the three men who signed the affidavit: Samuel G. Bailey, William. R. Handy, and Adam Knox.

1875 Artist Representation of Underwater Vessel

EARLY AMERICAN UFOS: 1800-1864

As to the second question, "Is it a sea monster?"[7] Frankly, the thing is too big for any type of sea monster -- the Blue Whale only reaches 100 feet itself and is earth's largest known life form alive today. So, due to its immense size and the aforementioned "hole" and "archway" on its shell-like body that resembled a clinker-built vessel, this thing would appear to be a craft. The men, in the year 1813, would no doubt have thought such a craft impossible to construct and so thought of it as a sea monster.

Something rather encouraging that lends this story some credence occurred in 2011. It was all thanks to the Swedish-based TV series *Ocean X Team*, a show about underwater treasure hunters. While in the Baltic Sea between Sweden and Finland, the team took an intriguing sonar image. The blurry image appears to be of a circular craft 200 feet in diameter with features atop it resembling ramps and stairways. When the hunters returned a year later to investigate further, a "mysterious electrical interference" prevented them from taking better photos and getting too close. Theories soon started popping up that the thing was a sunken UFO. Considering that the object was close to the size of the craft seen in 1813, perhaps it is some type of similar sunken craft? Still, today's scientists insist it is nothing more than an undersea rock formation.

---

[7] And yes, having an interest in Ufology, the authors are also well versed in cryptozoology, the study of hidden animals which includes sea serpents.

THE REAL COWBOYS & ALIENS

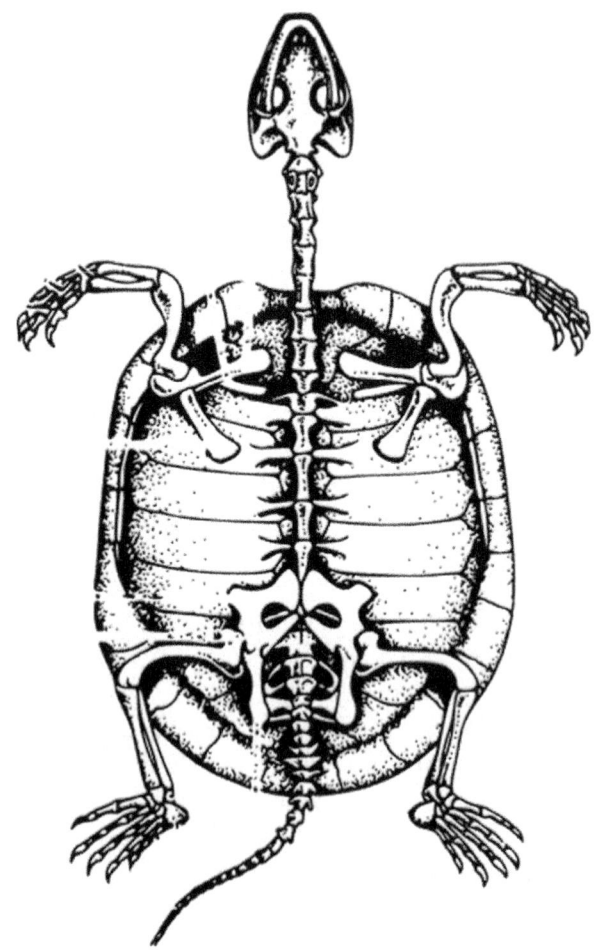

19th Century Illustration of a Turtle's Skeletal Structure

THE UFO LOOKED LIKE A TURTLE
July 25, 1813
Portsmouth, Virginia

IN THE PREVIOUS CHAPTER we discussed what was either a gigantic creature or craft with a shell-like structure for a body in the waters of the Atlantic. Oddly, only a few months later, another shell-like craft was spotted, this time in the skies of Portsmouth, Virginia. We say shell-like because at one point the witness likened it to a flying turtle!

But before we get to the actual story, we would like to take a detour to the most unlikely of spots: Tokyo, Japan in the 1960s. You see, there is the old adage that sometimes "Life Imitates Art." Specifically, it was Oscar Wilde who made the statement that "Life imitates art far more than art imitates life" in the opening of his 1889 essay "The Decay of Lying." But what about when art

THE REAL COWBOYS & ALIENS

unknowingly imitates life? Because, we're quite certain that a Japanese movie producer in mid-1960s Japan knew nothing about the Portsmouth UFO of 1813, which we will examine in this chapter.

Représentation de la Terre d'après les Hindous.

Drawing Representing the Belief that the Earth was Carried on Elephants Balanced on the Back of a Giant Turtle.

The producer we speak of was Masaichi Nagata, president of Daiei studios in the mid-1960s. Daiei was a direct competitor to Toho studios, whose Godzilla franchise had been going strong since 1954. In 1963 Daiei had planned their own giant monster movie to be called *Giant Horde Beast Nezura*. It was to be about giant rats that invade Tokyo. The ambitious production was locked in, and a huge miniature of Tokyo was constructed upon which real rats would be loosed.

EARLY AMERICAN UFOS: 1800-1864

When the rats were let loose upon the miniature of Tokyo, the action wasn't terribly exciting. The rats didn't do much of interest for the camera, though a few did start eating each other. Then the set became infested with fleas. With the set now a literal health code violation, the rats were all destroyed. Daiei was now left with a very expensive miniature of Tokyo, but no monster to destroy it.

Nagata was pondering this fact as he flew above the skies of Tokyo, and then it happened! Looking outside, he saw a cloud formation that looked just like a giant flying turtle![8] Then the idea hit Nagata to tell the story of a giant, flying turtle that attacks Japan.

It was at that moment that one of the more ludicrous Japanese monsters was born: Gamera, a giant flying and fire-breathing turtle - a friend to all children. Gamera's mode of flight was unique, as he would draw himself into his shell completely. After this, jets of fire would come from out of his arm and leg sockets, the shell would begin to spin, and Gamera would take to the skies in a manner similar to a flying saucer. Actually, Gamera's initial appearance causes observers to think that the monster was a flying saucer, which becomes a huge plot point of the movie.

The film, *Gamera*, was a big hit in Japan that inspired many sequels in which Gamera defended Japan from an array of bizarre monsters, like the giant vampire Gyaos. Later, the Gamera series became a favorite subject of *Mystery Science*

---

[8] Maybe it was a UFO hiding in a cloud? Nah.

THE REAL COWBOYS & ALIENS

Theater 3000, some might even say the giant turtle was their signature subject, considering they lampooned at least four of the Gamera films over the course of the show's run. How shocked would Joel and the bots be to find out that there was, in an abstract sense at least, a historical precedent for Gamera, the flying turtle?

Thomas Jefferson

As mentioned in chapter one, a report of the sighting of this bizarre apparition was submitted to Thomas Jefferson in 1813. The man who saw the fiery, turtle-like UFO was Edward Hansforth, who was a respected citizen, having served as the harbormaster for the District of Norfolk and Portsmouth in 1802.

On July 25, 1813, Hansforth and a friend, John L. Clark of Baltimore, saw something truly strange in the skies of Virginia. It was so puzzling, that Hansforth wrote a letter about it to Jefferson, who was then still president of the American

EARLY AMERICAN UFOS: 1800-1864

Philosophical Society. The letter is today preserved among The Thomas Jefferson Papers Series at the Library of Congress under the heading of General Correspondence, 1651-1827.

Original Letter from Hansforth to Thomas Jefferson

As to his strange encounter, Hansforth says he saw in the evening "in the South a ball of fire fully

THE REAL COWBOYS & ALIENS

as large as the sun at Meridian." The ball of fire emitted a great deal of smoke which caused the object to occasionally become hidden from the observers' view. Hansforth wrote that it was "frequently obscured within the space often minutes by a smoke emitted from its own body, but ultimately retained its brilliancy, and form during that period, but with apparent agitation."

And here's where it gets weird -- well, weird as far as UFO accounts go. "It then assumed the form of a turtle which also appeared much agitated and as frequently obscured by a similar smoke."

Now, what exactly Hansforth means by the appearance of a turtle is open to interpretation, as perhaps he meant to say the object was shaped like a turtle's shell? This would make the most sense, as a number of UFOs over the years seemed to be shell-like in appearance. In fact, a shell-shaped UFO was seen in the skies of Siberia recently in 2016. Petr Mironov, a witness, described it as a "kind of flame" in addition to being shell-like, much like the apparition Hansforth reported. Later that year, a similar shell-type craft was seen in Arizona as well.

In Hansforth's UFO from 1813, the object didn't retain its turtle shape for long. The UFO "descended obliquely to the West and raised again perpendicular to its original [height] which was on or about 75 degrees. It then formed the shape of a human skeleton." How it went from a turtle to a human skeleton is truly puzzling. Regardless of how this happened, after the transformation occurred, the object began to ascend and descend.

EARLY AMERICAN UFOS: 1800-1864

After a while, the mysterious thing then "disappeared within its own smoke."

As previously stated, the shell-like aspect makes sense in the context of UFO sightings, assuming Hansforth was talking about a turtle-shaped airship. However, its sudden transformation into the shape of a human skeleton is highly unusual in terms of recorded UFO sightings.

Was it a UFO or some kind of unidentified flying beast that falls squarely in the realm of cryptozoology? Could it possibly have been a creature thought to be long extinct? Given the lack of further information about this case, we will likely never know the answer.

THE REAL COWBOYS & ALIENS

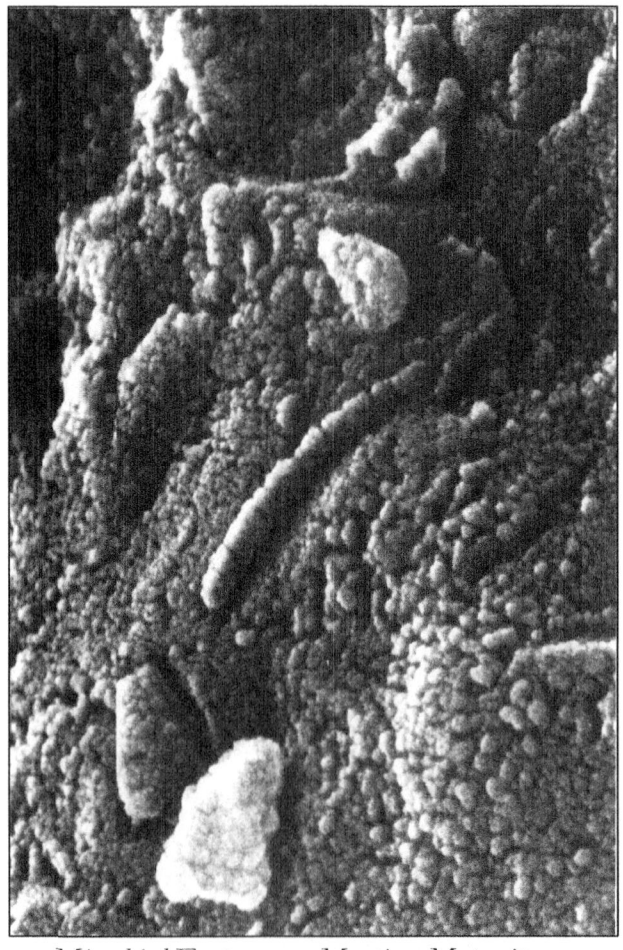

Microbial Features on Martian Meteorite
(NASA)

SLIME FROM THE STARS
August 13, 1819
Amherst, Massachusetts

STORIES OF SLIMY, gooey blobs that fall to Earth from outer space date back to the 1500s. In this chapter, we will explore a mysterious goo that fell at Amherst, Massachusetts in 1819.

More recently, on December 29, 1977, a Japanese research team discovered an ancient, Martian meteorite in the frozen Alan Hills region of Antarctica. Many years later, in 2019, it was revealed after a long study that the meteor contained trace amounts of fossilized organic material. Using powerful microscopes, a team of Hungarian scientists saw threadlike structures that are commonly associated with microbial features. In addition to that, mineralized remains of Martian

THE REAL COWBOYS & ALIENS

microbes could be seen. As such, this meteor fragment is now potential proof of former life on Mars.

Today's scientists apparently missed out on researching the gooey, meteoric remains found back in early 1800s, which they called "star rot" or "star jelly." This translucent to grayish white gelatinous substance has been found for years -- since the 14th Century in fact. John of Gaddesden, the English physician that wrote a treatise on medicine called the *Rosa Medicinæ* was the first to discover the strange substance which he termed *stella terrae* (Latin for "star of the earth" or "earth-star"). In his writings, Gaddesden described it as "a certain mucilaginous substance lying upon the earth" and went so far as to suggest that it could potentially be used to treat abscesses!

In the years to come, scientists would theorize that the substance was really terrestrial. They believed that it wasn't related to meteors at all and that the strange goo was simply the decomposed, leftover glands in the oviducts of dead frogs and toads that had come into contact with moisture as they rotted away in the elements. What was left over was the so-called star jelly.

However, many recorded cases of star jelly were too voluminous to have come from dead frogs, unless there were a great many dead frogs decomposing at once. *Fate* magazine put forth the theory instead that star rot was indeed cellular organic matter from the stars. In the rare instance that "star rot" is found today, it is sometimes explained away as a side effect of industrial waste in

EARLY AMERICAN UFOS: 1800-1864

addition to frog residue. In 2009 scientists commissioned by the *National Geographic* tested a star jelly sample but failed to find any sort of genetic material in the substance. Conversely, the BBC also did a test on a star jelly sample in 2015 and their results did point to it being leftover genetic material from a frog. From the differing results of these two studies we could perhaps deduce that sometimes star jelly simply is the decomposed, genetic material of frogs. But other times, the strange substance may have really been from the skies.

Amherst College, circa 1890 - 1901.

One of the best examples of this phenomena that did in all likelihood originate from the stars occurred in Amherst, Massachusetts on August 13, 1819. That night, between 8 and 9 p.m., a brilliant ball of white light resembling "burnished silver"

THE REAL COWBOYS & ALIENS

crashed into the ground. Actually, crash may not be the right word, and it may not have been an ordinary meteor. What makes this case special is that witnesses claimed that the ball of light descended slowly, not as rapidly as a typical meteor or shooting star.

Another View of Amherst College, circa 1903

A study published later in the *American Journal of Science* noted the following:

"[The meteor's] altitude, at its first discovery, was two or three times the height of the houses; it fell slowly in a perpendicular direction, emitting great light, till it appeared to strike the earth in front of the buildings, and was instantly extinguished, with a heavy explosion. At the same instant, as appeared from the report, and from the ringing of the church bell, an unusually white light was seen a few minutes afterwards by two ladies in a chamber of

EARLY AMERICAN UFOS: 1800-1864

Mr. Erastus Dewey. While they were sitting with two candles burning in the room, a bright luminous circular spot suddenly appeared on the side wall of the chamber near the upper floor in front of them, of the size of a two feet stand-table leaf. This spectrum descended slowly with a tremulous motion nearly to the lower floor and disappeared."

The next day, on the morning of August 14, the aforementioned Erastus Dewey walked out his front door to find a mass of strange goo twenty feet away in his yard. Upon learning of the incident, a chemist, Professor Rufus Graves, came onto the scene to investigate. Immediately he took interest in the details of the meteor's touchdown and the two women who had observed it from within Dewey's home. Of the room where the bright light was seen he wrote:

"In critically examining the chamber where the foregoing phenomenon was observed, it appeared that the light must have entered through the east front window in a diagonal direction, and impinged on the north wall of the chamber back of the ladies, and thence reflected to the south wall in front of them, forming the circular spectrum, with the corresponding tremulous motion of the meteor, and descending with it in the same direction, according to the fixed laws of incidence and reflection."

THE REAL COWBOYS & ALIENS

Portrayal of Professor Rufus Graves by an Actor in a Stage Play (Amherst College Archives)

As for the strange residue outside, Graves described it as resembling a saucer or plate face down on the ground. This is intriguing, as this is similar in shape to future reports of flying saucers! His report is worth reprinting in full:

"Early on the ensuing morning, was discovered in the door yard of the above-mentioned Erastus Dewey, at about twenty feet from the front of the house, a substance unlike anything before observed by anyone who saw it. The situation in which it was found, being exactly in the direction in which the luminous body was first seen, and in the

EARLY AMERICAN UFOS: 1800-1864

only position to have thrown its light into the chamber, (as before remarked,) leaves no reasonable doubt that the substance found was the residuum of the meteoric body.

"This substance when first seen by the writer was entire, no part of it having been removed. It was in a circular form, resembling a sauce or salad dish bottom upwards, about eight inches in diameter, and something more than one in thickness, of a bright buff color, with a fine nap upon it similar to that on milled cloth, which seemed to defend it from the action of the air. On removing the villous coat, a buff colored pulpy substance of the consistence of good soft soap, of an offensive, suffocating smell appeared; and on a near approach to it, or when immediately over it, the smell became almost insupportable, producing nausea and dizziness. A few minutes exposure to the atmosphere changed the buff into a livid color resembling venous blood."

Graves then placed a sample of the material in a half-pint tumbler. The chemist watched as the mysterious substance attracted moisture rapidly from there. It then began to liquefy into a "mucilaginous substance of the consistence, color, and feeling of starch when prepared for domestic use."

Then, sadly, the alien substance began to evaporate and within three days all that remained was a dark residue on the inside of the tumbler. Graves rubbed it between his fingers and watched it literally turn to ash. He then returned to the impact site where he had found the strange substance.

The place where the substance was first found was examined, and nothing was to be seen but a thin membranous substance adhering to the ground similar to that found on the glass.

Graves took samples of this substance as well and tested its reactions to various acids. "With the muriatic and nitric acids, both concentrated and diluted, no chemical action was observed, and the matter remained unchanged," he wrote. It was only when Graves experimented with a concentrated sulfuric acid that "a violent effervescence ensued." The substance began to dissolve and turn into a gas. Graves concluded, "There being no chemical apparatus at hand, the evolving gas was not preserved, or its properties examined."

Graves, who believed the substance to be something out of the ordinary, found himself embroiled in an argument with fellow chemist Edward Hitchcock of the Amherst College. Hitchcock was certain that it was simply a species of gelatinous fungus formed in the heat. Hitchcock felt the meteor sighting was merely a coincidence, while Graves insisted that the substance came from said meteorite.

Because this story came from the *American Journal of Science,* and not a random newspaper article, there was no doubt that Professor Rufus Graves was a real man. However, what the historical record reveals about him is surprising, and not particularly encouraging. Though he was something of a mover and shaker at Amherst

EARLY AMERICAN UFOS: 1800-1864

College -- in fact he helped raise the funds to get the college off the ground -- how he got his start as a professor is remarkable and lackluster at the same time.

As it turned out, Graves had been proprietor of a general store and the landlord to a chemist named Nathan Smith whom Graves had helped to find his first lecture room in 1797. Remarkably, Graves had merely been a former student that had attended possibly only one chemistry lecture when he was appointed as a professor of chemistry at Dartmouth College! In their book, *Improve, Perfect, & Perpetuate: Dr. Nathan Smith and Early American Medical Education*, authors Oliver S. Hayward and Constance E. Putnam took note of this strange turn of events and wrote on page 137 that, "Apart from attending one chemistry course, Graves's only claim to academic preferment -- his only academic involvement heretofore -- seems to have been that it was he who had found Smith his first lecture room in 1797." They also noted that, "Fortunately, perhaps, it appears that Graves may not have been called upon to teach the chemistry course very often..."

In another book by Putnam on the subject, *The Science We Have Loved and Taught: Dartmouth Medical School's First Two Centuries*, the author reveals that Graves was himself hired on as Nathan Smith's assistant in the chemistry lab. If this was before or after his appointment as professor is unclear, but Putnam wrote that it was not certain what made Smith think Graves was qualified for the position.

THE REAL COWBOYS & ALIENS

Graves's lack of credentials aside, it's too bad no traces of the substance he discovered remain today. Even if many cases of star jelly can be explained, it seems very well possible that the 1819 Massachusetts case might have been an exceptional example of the phenomenon, especially considering the nature of the strange meteor that seemed to slowly come down to earth.

In a strange postscript to this story, author Juanita Rose Violini, in her book *Almanac of the Infamous, the Incredible, and the Ignored*, writes that Graves's nemesis, Edward Hitchcock himself, said that "years later" there was another strange object that fell around the same exact spot, leaving behind the very same residue. Strange indeed!

10
ENCOUNTER AT MOUNT ADAMS
Circa 1816
Mount Adams, Washington

NAMED AFTER President John Adams, Mount Adams is a potentially active stratovolcano in the Cascade Range of Washington State. In fact, it is the second highest mountain in the state after Mount Rainier and is located only 34 miles east of that infamous volcano, Mount St. Helens.

Naturally, the mountain was first "discovered" by Lewis and Clark in 1805 as they travelled down the Columbia River. Of course, the Native Americans discovered the mountain first and called it both *Pahto* ("high up") and *Klickitat* ("beyond"). The natives there had many creation legends and myths regarding the mountain, but none seem to be extraterrestrial in nature (for cryptid fans, it was

THE REAL COWBOYS & ALIENS

said to be the home of a great Thunderbird, though).

Mount Adams c.1875
(Albert Bierstadt - Public Domain)

Back to Lewis and Clark, they noted that the formation was "a high mountain of immense height" and also supposed that it just might be "the highest pinnacle in America." Actually, they initially mistook the mountain for the already discovered Mount St. Helens, though on their return trip noticed both mountains and realized they were not one and the same. They did not bother to name the mountain, though. In fact, it wasn't officially designated as Mount Adams until 1853 when it was put on a map for the Pacific Railroad Survey. But we're getting ahead of ourselves as our tale takes place in the year 1816, after the mountain had been discovered but before it had been named.

EARLY AMERICAN UFOS: 1800-1864

Charles M. Skinner
(Brooklyn Daily Eagle)

The man responsible for unearthing the unearthly 1816 encounter was writer Charles M. Skinner. Actually, Skinner wasn't just a writer, he was the editor of the *Brooklyn Daily Eagle*. As such, he saw many a fantastic tale related in the paper's pages. Heck, he probably created some of them himself! Like many of us, Skinner had a desire to preserve the more unbelievable aspects of the country's history, that being tall tales and folktales. He published his favorites in a nine-volume set under the main title of *Myths and Legends of Our Own Land* in 1896. It is from the first volume that his stories on Mount Adams survive.

Some of the more interesting tales he glosses over in the lead up to the possible UFO sighting we will soon discuss included, "...talk of the discovery

THE REAL COWBOYS & ALIENS

of a magic stone; an Indian's skeleton that appeared in a speaking storm; [and] of a fortune-teller that set off on a midnight quest, far up among the crags and eyries." Before Skinner gets to the tale from 1816, some backstory and history is necessary regarding the French Indian War dating back to October 4, 1759. This was the day of the St. Francis Raid, a bloody part of the war near the southern shore of the Saint Lawrence River. On that day, Robert Rogers and 140 of his Rogers' Rangers, as they were called, had entered the village of St. Francis and slaughtered the inhabitants -- which reportedly mostly consisted of women, children, and the elderly. In the massacre, it is reported that anywhere from 30 to 300 people were killed in the village, while only one of Roger's men was killed and seven were wounded.

Drawing Depicting the Massacre at St. Francis

Six years later, a small detachment of nine of Rogers' Rangers were still making their way south back to America in the vicinity of Washington.

EARLY AMERICAN UFOS: 1800-1864

With them they carried looted treasures -- including a plate, candlesticks, and a silver statue -- that they had appropriated from the Church of St. Francis.

MAJOR ROBERT ROGERS.

However, in classic Western folktale tradition, all but one of the men would die on this return trip. In a turn of frontier justice, their Native American guide betrayed them when he led them among a "labyrinthine [of] gorges to the head of Israel's River, where he disappeared, after poisoning one of the troopers with a rattlesnake's fang."

One by one, the rangers succumbed to the cold and starvation and died off – only one survivor reaching a settlement, carrying a knapsack filled with human flesh. The treasure from the St. Francis church, namely the candlesticks, were

THE REAL COWBOYS & ALIENS

recovered near Lake Memphremagog, but the statue was never found.

It is at this point what we, the authors, consider a possible UFO sighting to begin, with what Skinner considers to be a ghost sighting. Skinner tells of a hunter camped at Mount Adams one night in the year 1816 where he saw a vision related to the dead Roger's Rangers.

Skinner writes, "The spirits of the famished men were wont, for many winters, to cry in the woods, and once a hunter, camped on the side of Mount Adams, was awakened at midnight by the notes of an organ. The mists were rolling off, and he found that he had gone to sleep near a mighty church of stone that shone in soft light. The doors were flung back, showing a tribe of Indians kneeling within. Candles sparkled on the altar, shooting their rays through clouds of incense, and the rocks shook with thunder-gusts of music. Suddenly church, lights, worshippers vanished, and from the mists came forth a line of uncouth forms, marching in silence. As they started to descend the mountain a silver image, floating in the air, spread a pair of gleaming pinions [the outer part of a bird's wing including the flight feathers] and took flight, disappearing in the chaos of battlemented rocks above."

Of course, the first sections all correspond to the story of Roger's Rangers -- the church, the massacre, etc. It is at the final point that this tale draws some similarities to UFO encounters when it mentions what appears to be a silver covered craft with pinion. Also worthy of mention is the "line of

EARLY AMERICAN UFOS: 1800-1864

uncouth forms" which could refer to some type of beings that emerged from the silvery UFO. Maybe the hunter saw these beings and later had difficulty recalling details in full, which often happens to persons who experience an alien abduction.

Perhaps the hunter describing this "vision" was in fact abducted at some point during the narrative. As mentioned, abductees are often unable to remember specific events and instead can offer only vague, general descriptions regarding what happened to them. Maybe he associated in his mind what he was seeing with the story of Rogers' Rangers, during the confusion surrounding the sudden appearance of bright lights and other disorienting images.

Whether it was a ghostly encounter or an alien one, this bizarre story is unique due to the history of the time in which it occurred.

THE REAL COWBOYS & ALIENS

From the Aurora and Pennsylvania Gazette.

Mr. Editor—A meteor of rather singular character made its appearance in the heavens last night, between the hours of eleven and twelve o'clock. As a great number of your readers, owing to the lateness of the hour, had not an opportunity of observing it, I shall endeavor to give some description of it.

It arose apparently from the neighborhood of the Schuylkill, and passing over Kensington and the river Delaware, finally disappeared behind the woods of Jersey. A long trail of light, like that of a shooting star was seen to follow it in the beginning of its ascension: large sparkles separated themselves from it and descending slowly, were distinctly visible until hidden from view by the tops of the houses. Its motions were rapid irregular, and wavering, like the fluttering of a kite or the rocking of an air balloon. Its appearance was of a deep red color, and remarkably brilliant; seemingly of about half the size of the moon. It arose until it crossed the Delaware, when it appeared but an inconsiderable speck scarcely discernable, and then descended with astonishing velocity until within a short distance of the horizon, where it remained stationary for a few moments. Suddenly it became exceedingly large and brilliant, sparkles again separated from the main body, and descended as before. It soon after became dim and disappeared behind the trees. Altogether, I should suppose it was visible about fifteen or twenty minutes.

New York Evening Post - July 17, 1829

11
THE UFO BEFORE KECKSBURG
July 16, 1829
Kensington, Pennsylvania

A STRANGE OBJECT crashing in a heavily wooded area near Kecksburg, Pennsylvania, became one of the best-known UFO cases in history in the year 1965. Incredibly, it was not unlike the descent of an unidentified object that was observed in the state of Pennsylvania 136 years earlier! In 1829, on the opposite end of the state from Kecksburg, another strange object seemed to descend into a wooded area across the Delaware River from Kensington, Pennsylvania, now a suburb of Philadelphia.

On the night of July 16, 1829, shortly before midnight, an unnamed observer saw an object streaking through the sky, describing it as "a meteor of rather singular character." The report of the

THE REAL COWBOYS & ALIENS

event appeared in New York's *The Evening Post* on the following day. No identifying information is given about the eyewitness, and the report contains no names of any persons.

Fireball Photographed in 2012 (Paola Castillo)

"A meteor of rather singular character made its appearance in the heavens last night, between the hours of 11 and 12. As a great number of your readers, owing to the lateness of the hour, had not an opportunity of observing it, I shall endeavor to give some description of it," the witness begins, without giving an explanation as to what he (or she) was doing at the time of the sighting. Regarding the reference to a "meteor," people of the nineteenth century had a limited vocabulary in terms of describing skyborne objects seen in the night sky. They knew nothing about aircraft, spacecraft, rocket ships, satellites, etc.

"It arose apparently from the neighborhood of the Schuylkill, and passing over Kensington and the river Delaware, finally disappeared behind the woods of Jersey." Given the directions mentioned,

EARLY AMERICAN UFOS: 1800-1864

to an observer in or near Kensington, the object would have come from the northwest and would have been heading to the southeast, across the Delaware River and into the state of New Jersey.

"A long trail of light, like that of a shooting star was seen to follow it in the beginning of its ascension; large sparkles separated themselves from it and descending slowly, were distinctly visible until hidden from view by the tops of the houses." Interestingly, unlike a meteor that would be descending from the upper atmosphere, this object was ascending as it moved toward the river from the northwest. As the object approached, the observer noticed large sparks of fire or electricity being emitted from the object and descending toward the earth.

"Its motions were rapid, irregular, and wavering, like the fluttering of a kite or the rocking of an air balloon." In addition to the fact that the object seemed to be ascending, not descending, it also had a strange "irregular, wavering, fluttering" sort of motion as it traversed the sky above the observer. In modern UFO sightings, the observed craft are often seen executing bizarre maneuvers like zigzagging and "flitting" around in the sky, which is what the Kensington object seemed to be doing. It was definitely not the smooth and steady downward streak of a meteor falling to the Earth.

"Its appearance was a deep red color, and remarkably brilliant; seemingly of about half the size of the moon." Although it is impossible to tell from the narrative how far the object was from the observer, his description is impressive. "Half the

THE REAL COWBOYS & ALIENS

size of the moon" would suggest that it was spherical and of considerable size. Its brilliant, deep red coloration could be associated with the heat and friction of entering the atmosphere.

"It arose until it crossed the Delaware when it appeared but an inconsiderable speck scarcely discernible, and then descended with astonishing velocity until within a short distance of the horizon, where it remained stationary for a few moments." The object seemed to continue rising until it crossed the Delaware River into Jersey, after which it moved downward "with astonishing velocity" and, from the observer's point of view, it rapidly diminished in size. An interesting observation is that the object, after approaching the horizon and on the verge of going out of view, suddenly stopped and remained stationary for a few moments, suggesting that it was under intelligent control. The rapid downward descent is also reminiscent of eyewitness accounts of the movement of UFOs in recent years.

"Suddenly it became exceedingly large and brilliant, sparkles again separated from the main body, and descended as before." After having come to a complete standstill in mid-air, the mysterious object seemed to grow in size and again began emitting an extremely bright glow. Sparks once again descended from the object down toward the ground."

"It soon after became dim and disappeared behind the trees. Altogether, I should suppose it was visible about 15 or 20 minutes." In the conclusion of this narrative, the observer notes that

EARLY AMERICAN UFOS: 1800-1864

the total duration of the sighting was between 15 and 20 minutes, which clearly rules out a free-falling meteor, as does the nature of its flight path and maneuvers. The object's behavior seems to clearly indicate that it was being controlled by an intelligence.

It is apparent that the object did not crash and thus did not create an impact crater or debris. No report of a crash or of the discovery of a meteorite was ever received. The object either made a controlled landing or changed course and departed the area rapidly. The length of the sighting -- 15 to 20 minutes -- again is incredibly significant in establishing that the object was not a meteor.

Artist's Rendering of the 1965 Kecksburg Object

In 1829, there were no air vehicles of any type that could account for this sighting. The first manned airplane flight was nearly 100 years in the future, and although hot air balloons did exist, they were not steerable, would not fly at night, and

THE REAL COWBOYS & ALIENS

would not emit light or sparks. The first steerable hot-air balloon, powered by a steam engine, made its maiden flight in 1852, but the technology was not practical. The first practical steerable balloons, powered by combustion engines, began to appear in 1898.

In addition to the *New York Evening Post*, this UFO sighting also appeared in *Hazard's Register of Pennsylvania: Devoted to the Preservation of Facts and Documents, and Every Kind of Useful Information Respecting the State of Pennsylvania*, Volume 4, page 48. A search by the authors of newspaper archives from 1829 found no other accounts of any reported meteors or meteor-like objects spotted in or around Pennsylvania.

> Mr. Editor:—A meteor of rather singular character made its appearance in the heavens last night, between the hours of eleven and twelve o'clock. As a great number of your readers, owing to the lateness of the hour, had not an opportunity of observing it, I shall endeavour to give some description of it.
> It arose apparently from the neighbourhood of the Schuylkill, and passing over Kensington and the river Delaware, finally disappeared behind the woods of Jersey. A long trail of light, like that of a shooting star was seen to follow it in the beginning of its ascension; large sparkles separated themselves from it and descended slowly, were distinctly visible until hidden from view by the tops of the houses. Its motions were rapid, irregular, and wavering, like the fluttering of a kite or the rocking of an air balloon. Its appearance was of a deep red colour, and remarkably brilliant, seemingly of about half the size of the moon. It arose until it crossed the Delaware; when it appeared but an inconsiderable speck scarcely discernable, and then descended with astonishing velocity until within a short distance of the horizon, where it remained stationary for a few moments. Suddenly it became exceedingly large and brilliant, sparkles again separated from the main body, and descended as before. It soon after became dim and disappeared behind the trees. Altogether, I should suppose it was visible about fifteen or twenty minutes.—*Aurora*.

Hazard's Register of Pennsylvania, Volume 4, page 48.

NIGHT OF THE LEONIDS!
November 13, 1833, North America

ON THE NIGHT of November 23, 1833, people throughout North America awoke to a beautiful yet terrifying sight, hundreds of meteors streaking across the skies unceasingly. We are not exaggerating; it has been estimated that around 72,000 or more shooting stars filled the sky per hour, or 20 per second.

It was astronomer Denison Olmsted, awakened by his neighbors, who observed that all the meteors were coming from the same origin point, the Constellation Leo, hence the eventual name Leonids. Immediately the next morning, Olmsted wrote to the newspapers about the origin of the phenomenon. The papers published his report, reassuring the masses that Judgement Day was not upon them.

*Astronomer Denison Olmsted
Discovered the Source of the Leonids*

Olmsted also encouraged his fellow citizens to write reports as to what they had seen, because, as he put it, "As the cause of 'Falling Stars' is not understood by meteorologists, it is desirable to collect all the facts attending this phenomenon, stated with as much precision as possible." And write they did, many of them to *The American Journal of Science and Arts*. This publication was run by Professor Benjamin Silliman, but in later years it was used as a reference by one Charles Fort. We, the authors, would be quite surprised if

EARLY AMERICAN UFOS: 1800-1864

you didn't know who that was, considering this book's subject matter. But, just in case you don't, Charles Fort was one of the first persons to conduct serious research into unexplained (possibly paranormal) phenomena. His research, and the strange news items that he collected, he called "the damned data" and thus wrote *The Book of the Damned* in 1919.

Charles Fort, Author of "The Book of the Damned"

And there was plenty of "damned data" to be collected from the Leonids meteor shower, even if it is forgotten by many serious researchers today. The incident was recorded in depth in *The American Journal of Science and Arts*, Volume 25 (and also carried over into Fort's better known of *The Book of the Damned*).

THE REAL COWBOYS & ALIENS

Illustration Depicting the Leonids Meteor Shower of 1833

EARLY AMERICAN UFOS: 1800-1864

In his very lengthy report in the *American Journal of Science and Arts*, Professor Amos Eaton listed three types of meteors, the first two basically being your typical run of the mill meteors and the other "fire balls." It is the third variety that we shall examine here, as they potentially relate to Ufology. The first of this strange category was seen in the city of Poland, in Trumbull County, Ohio. Eaton writes that, "A luminous body was distinctly visible in the north east, for more than an hour." Eaton continues that a Judge Calvin Pease saw it at five o'clock that morning near the star Alioth (Epsilon Ursae Majoris), the brightest star in Ursa Major. Pease's odd description given was "that it was then very brilliant in the form of a pruning hook, and apparently twenty feet long and eighteen inches broad, and that it gradually settled towards the horizon, until it disappeared." He implied that initially, to him, it resembled "a new moon, two or three hours high, shining through a cloud..." However, fifteen minutes later the heavenly specter vanished without a trace.

The strange, luminous body seen at Niagara Falls is even more interesting. Spotted at two o'clock, it was "a large luminous body, like a square table." The bizarre cloud, craft or whatever it was remained stationary for some time while emitting "large streams of light." This unusual formation was similar in shape to a cloud phenomenon spotted in Lewisburg, Virginia in 1864. In that case, a series of perpendicular, flat door-like clouds heralded the arrival of an unearthly army [you'll have to read the chapter to understand]. Another similar phenomenon, this time described as a coffin in the sky, was seen in Brownsville, Missouri, in 1875. Perhaps this was some sort of flat, rectangular craft hidden within a cloud?

The strangeness of that night didn't end with the two possible spacecraft; there was also plenty of "star jelly" to go around, too. A Mr. H. H. Garland of "the Nelson Company" wrote that he heard what he thought was a large drop of water fall on his roof. Garland went to investigate and "discovered a substance of about the circumference of a twenty-five-cent piece, of the consistence and appearance of the white of an egg made hot, or perhaps, animal jelly broken into fragments..."

The journal also described another large mass "of gelatinous matter" found at sunrise "which, from its singular texture, is supposed to have formed one of the large meteors." The journal said that it resembled soft soap and possessed "little elasticity." When heat was applied, it "evaporated as readily as water."

An unnamed woman milking a cow out in a field witnessed a sample of "star rot" fall to the ground right in front of her at sunrise. "On looking, she saw a round flattened mass, about a teacup or coffee cup full, looking like boiled starch, so clear that she could see the ground through it," the report said. At 10 a.m. she went to gather others to view the strange substance. Unfortunately, when she returned "no vestige of it remained." The report also states that, "A boy observed some minute white particles on the spot, as large as small shot, or pin's heads, of irregular shape, and falling

EARLY AMERICAN UFOS: 1800-1864

to powder, and disappearing when he went to take them up."

In addition to star rot, there was "fiery rain" which carried the star jelly down to the Earth with it. We also found this rather fascinating blurb on the fiery rain where one man wrote to a newspaper as to how he had witnessed "a globe of fire" explode in the heavens at four o'clock. He went on to write, "After this meteoric shower of fiery rain had for some time descended, a luminous serpentine figure was formed in the sky, which on its explosion, produced a shower of fire equally brilliant and incessant."

Back to the report from the *American Journal of Science*, the final strange item related, "One of our citizens was awakened by a ball of fire falling against his window."

There were other, totally unexplainable phenomena that occurred in conjunction with the meteor storm. The compiler noted that, "Near the time of the meteors, there were several remarkable events, which it may be well to record, although they may not have the least connection with the phenomenon under review."

A Dr. Kirtland, in Poland, Ohio, claimed that upon going to bed at 10 o'clock that night, he "discovered brilliant electrical sparks emitted from his clothes on any slight motion."

A correspondent of the *New York Daily Advertiser* claimed that the night after, on the evening of the 14th, while he was riding through Fredonia between six and seven o'clock "the tips of the ears of my horse, for a half an inch in length,

THE REAL COWBOYS & ALIENS

became luminous, and similar in appearance to phosphorescent bodies. It remained for some minutes."

There were strange, possibly unrelated phenomena preceding the meteor storm that were also reported. In Woodburn, near Hudson, on November 15th a man said, "A singular occurrence took place on my farm some days ago, which has excited a good deal of speculation among all who have visited the spot. A wood containing about an acre and a half, suddenly sunk down about thirty feet, most part of it perpendicularly; so that, where not long since the trees were to all appearance firmly imbedded, the topmost branches now peep out."

F. G. Smith of Lynchburg reported that the following occurred on November 13th: "Soon after 10 o'clock, I felt a slight repetition of the tremulous motion of the earth, which has repeatedly been observed in this vicinity of late." At Harvard, at about 8 o'clock in the morning that same day, reports indicated that "there was a slight shower of rain, when not a cloud was to be seen, the weather being what is called perfectly fair."

All this strange data confounded Professor Eaton to no end, especially the strange nature of the meteoric bodies. While Eaton was able to explain the different types of meteors thanks to their distances, he was naturally at a loss for the more "stationary bodies" such as the table formation over Niagara Falls. Of these he said, "The few remarkable bodies, which are described, as remaining for a long time stationary in a particular

EARLY AMERICAN UFOS: 1800-1864

part of the heavens, present anomalies which even conjecture is hardly competent to reach. We shall require more specific facts before we can attempt an explanation."

Modern scientists certainly attribute all of these reports to "fear and superstition," but perhaps that does not explain what happened. A lot of what was observed clearly was not related to meteor activity. It would be impossible for meteors to suddenly pause their downward course and hover in the sky. As the fictional detective Sherlock Holmes said, "When you have eliminated the impossible, whatever remains, however improbable, must be the truth.

THE REAL COWBOYS & ALIENS

"How Morning Star Lost Her Fish", from Stories the Iroquois Tell Their Children by Mabel Powers, 1917

13
LITTLE GREEN MEDICINE MEN
Circa 1858
Fort Bridger, Wyoming
and Utah Territory

OUR NEXT TALE takes us to Fort Bridger, Wyoming. The fort was named after its founder, the famous pioneer and fur trapper Jim Bridger, also called the "Daniel Boone of Wyoming." Bridger began the place as a fur trading post in 1842, on Blacks Fork of the Green River. Though the fort no longer exists today (being abandoned in 1890), it resided in what is now Uinta County, Wyoming. During its heyday it was a vital stopping point for wagon trains on the Oregon Trail, the California Trail, and the Mormon Trail. In 1858, due to tensions with Mormon settlers, who disliked the fact that Bridger sold liquor to Native Americans, the place became a military fort.

THE REAL COWBOYS & ALIENS

Jim Bridger

Long before that, in the early 1840s when Bridger settled the area, the place had some rather mysterious residents. They were called the Northern Paiutes, presumably named after the Paiute Indians. These Northern Paiutes were unlike the normal Paiutes, however, as they were described as a tribe of "small, smelly people." Reportedly, they only ate corn, which they received from the local natives. Odder still, they lived in holes in the ground but left the area upon the great influx of white settlers. Where did they go? Some said they may have travelled to Utah.

Though there are many "Little People" legends told by Native Americans, the little people spoken

EARLY AMERICAN UFOS: 1800-1864

of by the Utes in Utah sound incredibly similar to the ones recorded near Fort Bridger.

A study of the Utes conducted between 1914 and 1916 inadvertently ended up adding to the growing list of alien-like entities encountered by Native Americans. In this case, an entity that sounds strikingly similar to the *Star Wars* character Yoda served as a mentor to a Ute medicine man in the 1880s! This small creature sounds very similar in appearance to the beings seen in Wyoming.

Two Native Americans, Pa'gits and a Mrs. Washington, detailed various methods of healing the sick to the researchers. Both said that a medicine man would often have a "supernatural advisor." Pa'gits claimed that his mentor was a "little green man." He further elaborated that other medicine men were also under the tutelage of these small, green men and that there were many of them.

Pa'gits said he met his little green man when he was 12 years old. The being was two feet tall, green from head to toe, and carried a bow and arrow. Like Yoda, he was a "crotchety" teacher who was good to those he liked, especially medicine men. Those he didn't like, such as anyone that talked badly about him, he would pelt with arrows. The students he was training to become medicine men would then be paid by the victims to have the arrows removed!

Methods of teaching sometimes involved the little man singing to Pa'gits in his sleep. Hearing the songs in his sleep would help Pa'gits better remember their messages. Typically, if he wanted

THE REAL COWBOYS & ALIENS

to summon the little green man, he would wait for it outside facing east as he smoked a peace pipe before sunrise, since the little man only appeared at night. Other times he would leave a present near the front of the little man's home which coincidentally seems to resemble Yoda's hovel in the *Star Wars* movies. Pa'gits claimed that the homes of the little people looked like "little chimneys" scattered throughout the wilderness and other "unsettled" parts of the country.

Native American "Little People" from "Stories the Iroquois Tell Their Children" by Mabel Powers, 1917

At night the little men would light a fire. When passing any such dwelling, a passerby would be

EARLY AMERICAN UFOS: 1800-1864

obligated to leave a branch of cedar in front of the home as a respectful offering to the little men. Though not obligated to do so, upon successfully healing a sick patient, Pa'gits would often leave a small token or gift for his mentor.

Though it's unknown if George Lucas was aware of this specific story when he added the character to *Star Wars*, it is known that Lucas was inspired by tales of wizened little people. So there you have it, one of our favorite *Star Wars* characters seems to have his roots in the real life beliefs of Native Americans.

THE REAL COWBOYS & ALIENS

Meteor Streaking through the Sky

14
RETURN OF THE SPACE SLIME
May 1842, Dunkirk, New York

IMAGINE YOURSELF walking down the street one night, minding your own business, when suddenly above you streaks a massive meteor. But this is no typical meteor, nor is it a pleasant sight. It is accompanied by a strange, unsettling noise. And then, from out of nowhere, you are blinded by a bright light and then suddenly enveloped in a hot, smelly, slimy substance that burns your skin. This is exactly what happened to a Mr. Horace Palmer while walking home late one night in New York State.

Around 2:30 in the morning in early May of 1842, a "brilliant meteor" was observed by many citizens in and around Buffalo, New York. It was first seen southwest of the city at an elevation of 45

THE REAL COWBOYS & ALIENS

degrees heading north by the estimates of the *Buffalo Commercial Advertiser*. The large meteor then exploded somewhere over the northwest portion of the city "when it burst with a loud explosion, resembling the blasting of rocks, succeeded by a heavy rumbling sound for some moments." The paper continued, "The meteor was very large, and its light was like that of day. The sky was perfectly cloudless at the time. In the fore part of the evening there was quite a brilliant display of the aurora borealis."

> From the Baltimore Sun.
> *A Remarkable Meteor.*—The Chatauque county (N. Y.) Messenger, of the 13th instant, gives an account of an extraordinary meteor which appeared at that place, on the morning of the 11th instant, about three o'clock. Various observers decide it as being very large, and brilliant beyond example—that it made a grating or "whizzing" noise as it passed, and finally exploded with a report something like the sound of a distant cannon. At Dunkirk it was also noticed, and Mr. Palmer, whose veracity is vouched for, thus decides its appearance to him. We copy from the Messenger:

Portion of the Original Newspaper Story

Another firsthand account, by the editor of the *Westfield Messenger* published in the *Buffalo Unionist*, went as follows: "On Monday morning last about 3 o'clock, we were awakened by a sudden and extremely brilliant light which shone through

EARLY AMERICAN UFOS: 1800-1864

the window of our sleeping apartment. On opening our eyes, we had a momentary glimpse of a vividly luminous body or trail, which almost instantly passed out of sight, and was gone. We were convinced it was a large meteor and expected an explosion. We waited from three to five minutes, when a report burst through the welkin like a piece of heavy ordinance standing within a short distance. There was nothing in it like thunder, but a perfect resemblance to the sound we have named. It shook the house very sensibly, as it did others, and one instant jarring a toothbrush from the window to the floor. Its direction was northerly, and the explosion took place, probably, over the lake."

Dunkirk, N.Y., circa 1890-1900

"The light emitted was nearly as bright as daylight at Meridian. The shingles on distant houses were

distinctly visible. Mr. Sexton, our postmaster, was, at the time of its passage, sorting the mail having two candles to furnish him light; but the light of the meteor was so great as to make them appear like burning candles in full daylight."

While this was a fantastic meteor sighting to begin with, it gets even more so upon the entrance into the story of Horace Palmer. We shall resume the newspaper article, which itself is slightly skeptical of Palmer's account:

"The following rather startles our credulity, but we give to verbatim from the *Messenger*. Since writing of the above, we have conversed with Mr. Horace Palmer, who was on his way from Dunkirk to this place, when the meteor appeared. He was two or three miles from Dunkirk, when he appeared to be instantly surrounded with the most painful vivid light proceeding from a mass of fluoride or jelly-like substance which fell around and upon him, producing a sulfurous smell, a great difficulty in breathing, and the feeling of faintness, with a strong sensation of heat."

"As soon as he could recover from his astonishment, he perceived the body of the meteor passing above him, seeming to be about a mile high. It then appeared to be in diameter about the size of a large steamboat pipe, near a mile in length. Its dimension varied soon; becoming first much broader, and then weaning away in diameter and length until the former was reduced to about 8 inches, and the latter to a fourth of a mile, when it separated into pieces which fell to the earth, and

EARLY AMERICAN UFOS: 1800-1864

almost immediately he heard the explosion, which, he says, was tremendous."

It is interesting to note that Mr. Palmer perceived the meteor as changing in size. One has to wonder if he was disoriented from the slime. Or, was he inebriated, considering that he was walking home so late at night? Another possibility is that the meteor wasn't actually a meteor, but an amorphous mass of slime falling from space.

Returning to the story of Palmer, the account continues, "On arriving home in the morning his face had every appearance of having been severely scorched; his eyes were much affected, and he did not recover from the shock it gave his system for two or three days."

"This really is a marvelous story, but Mr. Palmer is a temperate and industrious man, and a man of integrity, and we believe that anyone conversing with him on the subject, would be satisfied that he intends no deception, but describes the scene as nearly as possible, as it actually appeared. Probably, however, his agitation at his sudden introduction to such a scene, caused the meteor to be somewhat magnified to him. Witnesses here speak of the sparks which were given off; probably one of these fell and enveloped Mr. Palmer. In addition to its light, Mr. Palmer states that its passage was accompanied by a sound like that of a car moving on a railroad, only louder."

In truth, one shouldn't be able to hear a meteor because it is so far away. However, scientists at Sandia National Labs led by Richard Spalding conducted a study on this phenomenon. They

THE REAL COWBOYS & ALIENS

theorized that since fireballs sometimes emit bright blasts of light, these blasts can actually heat the surfaces of objects miles away. The sudden change in temperature then produces sound.

Jelly-like Substance Possibly from Meteorite By James Lindsey at Ecology of Commanster
[CC BY-SA 2.5 (https://creativecommons.org/licenses/by-sa/2.5)]

In the *Scientific Reports* journal, the team explained it like this: "We suggest that each pulse of light can heat the surfaces of natural dielectric transducers. The surfaces rapidly warm and conduct heat into the nearby air, generating pressure waves. A succession of light-pulse-produced pressure waves can then manifest as sound to a nearby observer."[9]

This then, could have been the source of the noises that witnesses heard in connection to the meteorite -- unless it was some sort of spacecraft that is.

---

[9] https://www.livescience.com/57783-weird-cause-of-meteor-noises-found.html

EARLY AMERICAN UFOS: 1800-1864

And then there's the slime, the so-called Star Rot or Star Jelly, whichever you prefer. As we established in a previous chapter, to this day scientists still aren't entirely sure what it is, or if it really does come from meteors. But, for a modern-day example, one need look no further than the Russian Chelyabinsk meteor strike of 2013. Right after it occurred, a strange greenish astral jelly began appearing in a wildlife park in Somerset, which some thought might be linked to the meteor.

2013 Chelyabinsk Meteor Trail by Alex Alishevskikh
[CC BY-SA 2.0 (https://creativecommons.org/licenses/by-sa/2.0)]

Thus ends the account of the enigmatic Mr. Horace Palmer. Unfortunately, we were unable to turn up any records of his existence, though that doesn't mean he didn't exist. It simply would have been encouraging to find any further stories on this man, what happened to him after this strange incident, etc. Either way though, as far as we know, Palmer is the only individual in existence to have been "slimed" by a meteor as it passed by overhead.

THE REAL COWBOYS & ALIENS

Wonderful Phenomena.

Singular appearance of the moon—a black spot on the surface—the moon split into fragments—lights shooting off from it and forming into the shape of a man!

Mr. James D. King a respectable citizen of this county, and a gentleman of undoubted veracity, called at our office on Wednesday and gave the following details of a most remarkble appearance of the moon, for about an hour between seven and nine o'clock on Tuesday night last.

He states that being in the habit of noticing the appearances of the moon at this season of the year with a view to the common prognostication of whether it would be "wet or dry," he observed, while looking with that object on Tuesday night, that it appeared at first much larger than common, nearly three times the usual size, and more like a circular sheet of fire than like an ordinary moon. In a few moments a very dark and deep black spot was plainly discernible about the centre of the moon's disc, which immediately commenced playing up and down, backwards and forwards on the surface, and as the spot approached the upper edge it grew less, and a faint light distinctly shone through it. This spot finally became stationary in the centre, when the moon divided into three seperate fragments, each giving distinct and seperate lights, being of irregular forms and appearing as though the spot had split them off. Then the moon gradually returned to its original appearance, and from that again looked naturally.

What he had already seen was so remarkable that Mr. King, with his family, continued the observation, and but a few moments had elapsed before the black spot again appeared, and again the moon divided—this time into four distinct, irregular parts or fragments. And immediately a light resembling the tail of a comet shot from the lower fragment at the south-east corner, apparently some three or four feet downwards, while another much larger, from the upper portion or north-west corner, struck off directly upwards, to the length of between five and six feet.

May 6, 1843 issue of the Panola, Mississippi newspaper, The Weekly Register

15

THAT'S NO MOON!
May 2, 1843
Panola County, Mississippi

IN RECENT DECADES, some of the world's most fascinating images of unidentified flying objects have been taken by observers who aim their cameras at the moon and capture objects moving across the moon's luminous surface. Many years before cameras were readily available, in 1843, the moon became the backdrop for a spectacular UFO sighting by amateur astronomer James D. King in Northern Mississippi. King, who had gone outside at 9 p.m. to observe the inclination of the crescent moon relative to the Earth's horizon, was described by the local newspaper as "a respectable citizen of this county, and a gentleman of undoubted veracity." King was not alone in his observation; also witnessing the spectacle was his wife, 13-year-old daughter, and an unnamed "young lady," all of whom swore that what they reported was accurate.

Photo of the Moon (Public Domain)

On May 2, 1843, in Panola County, Mississippi, King was observing the moon in an effort to determine whether it was in a "wet moon" or "dry moon" configuration, which has to do with what angle the crescent moon is to the horizon. In either case, King was expecting to see a crescent moon, not a full moon, and certainly not the object that he saw when he looked up at where he expected the crescent moon to be.

For an hour, King and the others with him were amazed to see in the sky where he expected to see the moon, a huge orb, fully three times the size of the moon. Instead of the familiar features of the moon, he saw what looked like a "circular sheet of fire." It seems clear that this unidentified object was interposed between him and the moon and was

EARLY AMERICAN UFOS: 1800-1864

obviously not really the moon at all. So unusual was the sight that he decided to make his observation public, and it was published in the May 6, 1843 issue of the Panola, Mississippi newspaper, *The Weekly Register* and was also carried in the *New York Tribune*, the *Mississippi Free Trader* of Natchez, Mississippi, and other newspapers of the time.

As King watched this strange object blocking his view of the moon, he noticed some activity from within the surface of the orb. He later told a reporter, "In a few moments, a very dark and deep black spot was plainly discernible about the center of the moon's disc, which immediately commenced playing up and down, backwards, and forwards on the surface, and as the spot approached the upper edge, it grew less, and a faint light distinctly shown through it."

King's description suggests that he was observing a series of apertures or portals that slowly opened along the center of the orb, distorting its surface as it opened. The "faint light" that "shown through it" was clearly a source of illumination from within the object itself, shining through the newly opened doors or hatches in the object's hull.

"This spot finally became stationary in the center, when the moon divided into three separate fragments, each giving distinct and separate lights, being of irregular forms and appearing as though the spot had split them off."

It seems clear that three separate windows or hatches had opened in the hull of this immense UFO. Shortly after the opening of these apertures,

THE REAL COWBOYS & ALIENS

they once again closed, and King said that "the moon gradually returned to its original appearance, and front that again looked naturally. Moments later, the apertures opened again, only this time there were four instead of three, which King referred to as northwest (upper left), northeast (upper right), southwest (lower left), and southeast (lower right).

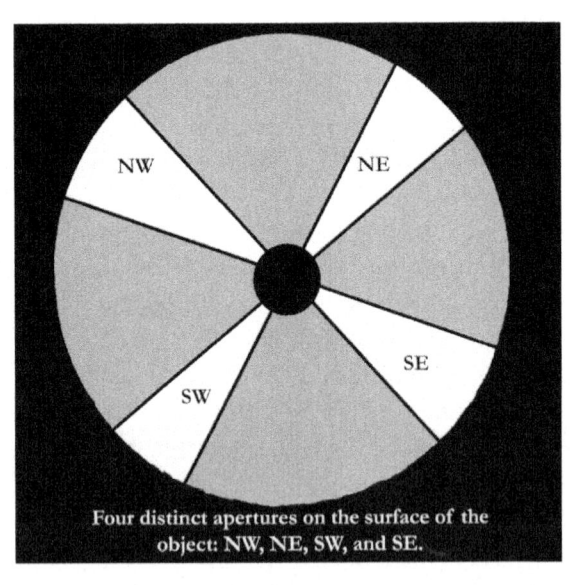

Four distinct apertures on the surface of the object: NW, NE, SW, and SE.

Out of the southeast aperture, a smaller object suddenly shot out, which King described as follows: "... immediately a light resembling the tail of a comet shot from the lower fragment at the southeast corner, apparently some three or four feet downwards." Clearly something was launched from the main UFO, perhaps a smaller scout ship.

EARLY AMERICAN UFOS: 1800-1864

But the most remarkable sight was yet to come. King observed a second, much larger, object come out of the northwest corner and move slightly upward from the main ship.

King said, "... another much larger [object], from the upper portion or northwest corner, struck off directly upwards, to the length of between five and six feet. This last now went off and left the corner apparently four feet or more and formed into the shape of a man—standing erect. The figure was of the most perfect imaginable symmetry, of about the medium size and height, clothed in the purest snow white, and the back alone presenting itself to view. It was visible a few moments, when gradually the figure changed to the simple light."

This second object took the form of a humanoid, with its back turned to King and bathed in an intense white light, which after a few moments seemed to consume the figure, leaving only light where the human shape had been. The description almost sounds like an occupant of the UFO was transitioning between states of matter or was being converted into energy as a means of transitioning between dimensions or between physical location -- a "transportation beam" perhaps.

After this, the two smaller orbs of light joined back into the larger sphere, and the entire object returned to its original appearance. King said, "... the lights retreated to the fragments, these again came together, and the moon resumed a natural appearance."

Following this astonishing sighting, King was uncertain whether to make it known to the public

THE REAL COWBOYS & ALIENS

or not. The reporter that later took his story down said, "He protests that in calling on us to make public these facts, he has no motive but to tell a plain, unvarnished tale of truth, and leaves others to judge of its import -- that he was not in the least alarmed or agitated but is much in his sober senses as he ever was in his life and in order that no one should have occasion to doubt the sincerity of his narrative, he has authorized us to give his name and to refer to his family as witnesses of the scene with himself. To what causes it is attributable, he does not know, whether it was an optical illusion affecting his whole family at once -- an operation of nature never before witnessed, or something of a miraculous character, he does not undertake to determine, but avows that this statement from which we have deviated, if at all, in no essential particular, is true and will at all times be maintained to be true on his honor and character as a man, as he will convince any more fully who may choose to inquire of him further in relation to it."

The newspaper article also assured the public, "The family of Mr. King, consisting of his wife and a daughter, 13 years of age, with another young lady, all witnessed what is above related."

Another interesting angle to this sighting is that it attracted the attention of a group of religious zealots known as the Millerites, who believed that the Second Coming or "Advent" of Jesus Christ would take place in 1843 or 1844 and were keenly focused on "signs and wonders" appearing in the sky. The Millerite movement, known as *Millerism*, was founded on the teachings of William Miller, who in 1833 began preaching about the impending Second Advent.

William Miller (Public Domain)

In *Unnatural Phenomena: A Guide to the Bizarre Wonders of North America,* author Jerome Clark notes that the story of James D. King was thought to be of interest to the Millerites, and a correspondent, in quoting from the story, wrote, "As this account may interest some of our Millerite friends, we make an extract from it."

As signs and wonders go, the sighting of a monstrous UFO by James D. King was a doozy. The Millerites, who were already looking for some sign of the impending Second Coming, no doubt were extremely interested to hear of King's observation. In its writings, the sect often cited the appearance of comets, meteors, and other celestial phenomena as being portends of the coming of Jesus Christ. Out of the beliefs of William Miller arose a number of current Christian denominations, including the Seventh-day Adventist Church. Also influenced by Miller's teaching was the sect that later became the Jehovah's Witnesses.

Perhaps Miller was not totally incorrect about a monumental event happening in 1843 that involved humanity having contact with an extraterrestrial intelligence? Perhaps the event was what was witnessed by James D. King on May 2, 1843, in Panola County, Mississippi?

16
ZAPPED BY A UFO
November 2, 1851
Schenectady, New York

OVER THE YEARS since the 1940s, eyewitnesses have seen UFOs that seem to use clouds to remain hidden from view as they traverse the skies over our cities. Sometimes it appears that these mysterious craft generate a cloud-like vapor that surrounds the vessel. Other times, they simply use existing clouds to hide behind, only darting out when necessary.

As we have established, terms like "UFO," "rocket ship," "spaceship," and "flying saucer" were not in the vocabularies of the people of the nineteenth century. Instead, they referred to unknown celestial objects as meteors, comets, etc. Such was a case that appeared in the *New York*

THE REAL COWBOYS & ALIENS

Times, on November 6, 1851, in which an intensely blueish-white orb that witnesses called "a very singular meteor" zoomed out of a cloud and emitted some kind of energy beam, striking a young man on the ground.

> —A very singular meteor was observed passing over the west part of the city of Schenectady, on Sunday evening. It appeared to emit itself perpendicularly from a dark cloud, expand and contract in size, and withdraw again in the cloud. The color was an intense blueish white. A young gentleman directly beneath it, received a sharp electric shock. No report of an explosion was heard.

The New York Times, 11-6-1851, p. 3

The story was carried in a number of the nation's newspapers. The incident happened on Sunday, November 2, 1851, over the western part of the city of Schenectady, New York. It began when observers noticed a strange looking dark cloud in the sky above them.

As they gazed at the cloud, an object suddenly darted out of the cloud bathed in an intense blueish-white glow. Another interesting feature of the object is that it seemed to "expand and contract in size," indicating that the strange orb was pulsating or throbbing as it moved out of the cloud. This could indicate that it was "powering up" with some type of energy, preparing to discharge or utilize the energy somehow.

As it turned out, the object's energy was directed downward toward a "young gentleman" who was observing it from the ground "directly beneath it," and the results were surprising. The young man

EARLY AMERICAN UFOS: 1800-1864

"received a sharp electric shock." Interestingly, although the witness was struck by the energy, no sound was heard, such as would indicate that an explosion had occurred. Also, the other witnesses said that after the energy discharge struck the young man below, the glowing orb then "withdrew again in the cloud."

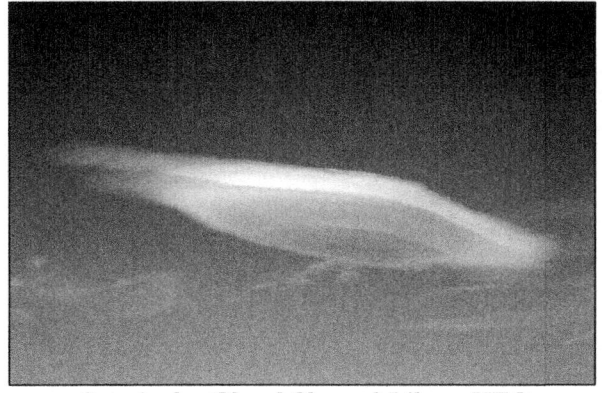

Lenticular Cloud Shaped Like a UFO
[CC BY-SA 3.0 (http://creativecommons.org/licenses/by-sa/3.0/)]

Unfortunately, the historical record does not give the name of the victim of this incident; nor does it state how badly he was injured. The phrase "a sharp electric shock" suggests that it did not result in serious injury, but rather was simply a relatively mild electrical shock.

Stories abound in UFO literature about witnesses being injured as a result of energies emitted from a UFO. In book two of this series, we tell the story of a cowboy who was seriously injured when he ventured too close to a UFO that crash-landed near Benkleman, Nebraska on June 6, 1884, as

THE REAL COWBOYS & ALIENS

reported in the *Nebraska Nugget* and the *Daily State Journal* of Lincoln, Nebraska.

> A METEOR.—A very singular meteor was observed passing over the west part of the city of Schenectady, N. Y., on Sunday evening. It appeared to emit itself perpendicularly from a dark cloud, expand and contract in size, and withdrew again in the cloud. The color was an intense blueish white. A young gentleman directly beneath it, received a sharp electric shock. No report of an explosion was heard.

The Perry County Democrat (Bloomfield, PA), 11-13-1851, p. 2

In more recent years, the best-known case of a beam of energy striking an observer on the ground happened in 1975 in a Northern Arizona forest, when forestry worker Travis Walton encountered a large, glowing UFO. Looking up at it from directly underneath, what happened next is described by Walton in his book *Fire in the Sky*:

"I ducked into a crouch when a tremendously bright, blue-green ray shot from the bottom of the craft. I saw and heard nothing. All I felt was the numbing force of a blow that felt like a high-voltage electrocution. The intense bolt made a sharp cracking, or popping, sound. The stunning concussion of the foot-wide beam struck me full in the head and chest. My mind sank quickly into

EARLY AMERICAN UFOS: 1800-1864

unfeeling blackness. I didn't even see what hit me; but from the instant I felt that paralyzing blow, I did not see, hear, or feel anything more."

"The men in the truck [his six companions] saw my body arch backward, arms and legs outstretched, as the force of the blow lifted me off the ground. I was hurled backward through the air ten feet. They saw my right shoulder hit the hard-rocky earth of the ridgetop. My body landed limply and lay motionless, spread out on the ground."

Travis Walton Speaks About His Encounter (2018 Photo by Noe Torres)

In private conversations with author Noe Torres, Travis Walton has stated that he does not think the UFO intentionally directed an energy beam at him in an attempt to hurt him. "I probably was just standing in the wrong place at the wrong time, as the ship was beginning to power up its propulsion system," Walton said. He added that before being blasted, he felt the air around him suddenly become "charged up," as if by static electricity. He felt as though an electrical charge was building up in the environment around him, followed by the discharge that impacted him.

Since Travis Walton's case is one of the world's best-known UFO abduction stories, it is fascinating to find a similar story from way back in 1851 in Schenectady, New York. The similarities between Walton's case and the 1851 incident are striking indeed.

17
UFOS OVER CAMPUS
June 1, 1853, Spencer, Tennessee

UFO SIGHTINGS with multiple witnesses at institutions of higher learning are extremely rare, even in our day, which makes an incident that happened in 1853 at Burritt College in Spencer, Tennessee, such a treasure trove to modern UFO researchers. It is considered one of the most significant UFO sightings of the nineteenth century in North America. Both faculty and students witnessed the sighting of two objects over the college, and Professor A. C. Carnes wrote a report about it that was published in the July 2, 1853 edition of *Scientific American*. As such, this event is one of the best documented multiple witness UFO sightings of the nineteenth century.

Singular Phenomenon.

We have received a letter from Professor A. C. Carnes, of Burritt College, Tenn., with the following account of a singular phenomenon, that was seen by a number of the students, on June 1st., at 4½ A. M., just as the sun was rising:—

"Two luminous spots were seen, one about 2° north of the sun, and the other about 30 minutes further in the same direction. When seen, the first had the appearance of a small new moon; the other that of a large star.— The small one soon diminished, and became invisible; the other assumed a globular shape, and then elongated parallel with the horizon. The first then became visible again, and increased rapidly in size, while the other diminished, and the two spots kept changing thus for about half an hour. There was considerable wind at the time, and light fleecy clouds passed by, showing the lights to be confined to one place."

The students have asked for an explanation, but neither the President nor Professors are satisfied as to the character of the lights, but think that electricity has something to do with it. The phenomenon was certainly not an electrical one, so far we can judge, and possibly was produced by distant clouds of moisture.

Scientific American, July 2, 1853, pg. 333.

EARLY AMERICAN UFOS: 1800-1864

The sighting began at 4:30 a.m. on June 1, 1853, "just as the sun was rising," as a number of the students observed two unusual, glowing objects in the sky. Burritt College (1849-1939) was one of the first coed institutions in the South, and one of the first state-chartered schools in south-central Tennessee. Affiliated with the Churches of Christ, the college offered a classical curriculum, and stressed adherence to a strict moral and religious code. A typical school day began at 5 a.m., with students preparing their rooms for inspection. This was followed by a one-hour study period and a half-hour devotional in the main hall. Classes were held throughout the remainder of the day.

So early on June 1, the students were preparing for inspection of their rooms when they noticed the anomalous sight in the sky over the college. They were soon joined by A. C. Carnes, professor of mathematics, who documented the sighting in fairly precise detail and later submitted a report about it to *Scientific American*.

"Two luminous spots were seen, one at about 2 degrees north of the sun, and the other about 30 minutes further in the same direction." In the summer of 1839, the sun rose in the east-northeast. Holding one's little finger out at arm's length represents about 1 degree; so, Carnes wrote that the object was about two little fingers north of the rising sun, right along the horizon. Carnes added that this first object, which was the larger of the two that were observed, had the appearance of "a small new moon." This suggests that the object was round or spherical. At a later point, Carnes said that the object "assumed a globular shape, and then elongated parallel with the horizon."

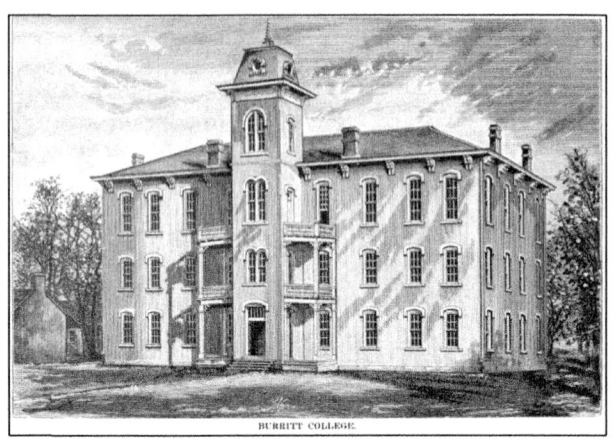

Photo of Burritt College from "Burritt, our alma mater" by Effie Gillentine-Ramsey and William Billingsley, 1914.

The assumption is that the spherical shape of the moon-like object became more clearly defined, but it then seemed to "elongate," presumably into a cylindrical shape. This suggests the possibility that the observers were at first looking at the round end of a cylinder, and later, the cylinder rotated horizontally, causing it to appear elongated, parallel with the horizon.

"When seen, the first had the appearance of a small new moon; the other that of a large star. – The small one soon diminished and became invisible; the other assumed a globular shape and then elongated parallel with the horizon." Just slightly to the left of the moon-like object was a second "spot" that looked more like a "large star."

EARLY AMERICAN UFOS: 1800-1864

Its relationship to the moon-like object is unclear, but perhaps it was a wave, glow, light, or reflection of light coming from the moon-like object. It seems likely that this was the case, since the two "spots" were in very close proximity to each other.

Two Orbs in Space, in this Case Being the Moons of Mars (NASA Photo)

"The first then became visible again, and increased rapidly in size, while the other diminished, and the two spots kept changing thus for about half an hour." Again, this description suggests that the larger, moon-like object was cylindrical in shape and was rotating or spinning; whereas the object alongside it was closely associated with it, perhaps a light or wave emanating from it. As the larger object moved, the smaller object would either appear or disappear. The rapid increase in size of the first object seems to indicate that it was moving closer to the observers.

"There was considerable wind at the time, and the light fleecy clouds passed by, showing the lights to be confined to one place." In this comment,

THE REAL COWBOYS & ALIENS

Professor Carnes is pointing out that these objects were not extremely distant, as in outer space, and that they remained fairly stationary in their location in the sky during the observation. The total length of the sighting was at least 30 minutes and possibly longer.

Entrance to the former Burritt College in Spencer, Tennessee (Courtesy Brian Stansberry via Wikipedia)

The reporter for *Scientific American* adds, "The students have asked for an explanation, but neither the President nor Professors are satisfied as to the character of the lights but think that electricity has something to do with it. The phenomenon was certainly not an electrical one, so far we can judge, and possibly was produced by distant clouds of moisture." Obviously, there was a struggle to attach a rational, scientific explanation to this highly

EARLY AMERICAN UFOS: 1800-1864

unusual sighting but without much success. No mention of electrical phenomena occurs in Professor Carnes' original observation.

This case has been mentioned in many books that compile unexplained phenomena throughout American history, one of the most recent being Jerome Clark's 2005 book, *Unnatural Phenomena: A Guide to the Bizarre Wonders of North America.* It is also included in an account of the most significant UFO sightings of the nineteenth century by the web site *HowStuffWorks.com,* which comments, "Unspectacular though it was, the event was certainly a UFO sighting, the type of sighting that could easily occur today. It represented a new phenomenon astronomers and lay observers were starting to notice with greater frequency in the Earth's atmosphere. And some of these sights were startling indeed."

> Rev. W. D. Carnes, of Burritt College, Van Buren County, Tenn., was elected president on the 20th of March, 1858, and at once accepted. The new president was a Christian minister and an alumnus of the university, having graduated in 1842. He was tutor in 1842-43 and principal of the preparatory department from 1843 to 1848. At a later date the faculty was completed as follows: M. C. Butler, ancient languages and literature; A. C. Carnes, mathematics, and Rev. John Washburn, principal of the preparatory department. Tuition was put at $25 in college and $20 in the preparatory department for the term of five months. The president received from the endowment fund $400 and each of his assistants $250. Their salaries were increased by a pro rata of all tuition fees.

SOURCE: "Higher Education in Tennessee." Circular of Information of the Bureau of Education, No. 5, 1893

THE REAL COWBOYS & ALIENS

Crumbs for the Curious.

A WILD MAN IN MAINE.—The following communication relative to the discovery of a wild man in Waldoboro' was handed to the editor of the Thomason *Journal*, by Mr. J. W. McHenri. The *Journal* says there are several persons who can vouch for the truth of the statements which it contains:

Mr. Editor: On the morning of January 2d, while engaged in chopping wood a short distance from my house in Waldoboro', I was started by the most terrific scream that ever greeted my ears; it seemed to proceed from the woods near by. I immediately commenced searching round for the cause of this unearthly noise, but after a half hours fruitless search I resumed my labors, but had scarcely struck a blow with my axe when the sharp shriek burst upon the air. Looking up quickly, I discovered an object about ten rods from me, standing between two trees, which had the appearance of a minature human being. I advanced towards it, but the little creature fled as I neared it. I gave chase and after running some distance succeeded in catching it.

The little fellow turned a most imploring look upon me, and then uttered a very sharp, shrill shriek, resembling the whistle of an engine. I took him to my house and tried to induce him to eat some meat, but failed in the attempt.— I then offered him some water, of which he drank a small quantity. I next gave him some dried beach nuts, which he cracked and ate readily. He is of the male sex, about eighteen inches in height, and his limbs are in perfect proportion. With the exception of his face, hands and feet, he is covered with hair of a jet black hue. Whoever may wish to see this strange specimen of human nature, can gratify their curiosity by calling at my house, in the eastern part of Waldoboro', near the Trowbridge tavern I give these facts to the public, to see if there is any one who can account for this wonderful phenomenon. J. W. MCHENRI.

Original Article in the Brooklyn (New York) Daily Eagle — Jan. 13, 1855

THE 18-INCH HUMANOID

January 2, 1855
Waldoboro, Maine

VENTURING ONCE MORE to the state of Maine, which is consistently in the top five of all U.S. states for UFO sightings per capita, a very strange incident happened in 1855 less than 20 miles from where schoolteacher Cynthia Everett spotted a luminous UFO in 1808. In this case, an unearthly, small humanoid referred to as a "wild man" was found in the woods of Waldoboro, and was captured, in an incident that received wide coverage in the news media of the day and has appeared in numerous publications about unexplained events occurring in the nineteenth century. Originally, the story titled "A Wild Man in Maine," appeared in the *Thomason (Maine)*

THE REAL COWBOYS & ALIENS

Journal, from where it spread to *The Brooklyn (N.Y.) Daily Eagle*; *The Kentucky Tribune*; *The Daily Journal* of Wilmington, North Carolina; *Hornellsville (N. Y.) Weekly Tribune*; and many other publications of 1855.

The episode began on the morning of January 2, 1855 in Waldoboro, Maine, when Mr. J. W. McHenri (possibly an alternate spelling of McHenry) encountered the strange creature while chopping wood a short distance away from his house. At the time of the sighting, Waldoboro, the county seat of Lincoln County, had a population of 4,199 people, according to the 1850 U.S. Census. Known for its strong agriculture, fishing, and shipbuilding, Waldoboro was the location where the world's first five-masted schooner was launched in 1888.

The Schooner Governor Ames Preparing for Launch, Leavitt-Storer Shipyard, Waldeboro, Maine (Wikipedia)

EARLY AMERICAN UFOS: 1800-1864

Illustration by Morburre

[CC BY-SA 3.0 (https://creativecommons.org/licenses/by-sa/3.0)]

McHenri was chopping wood when he was suddenly startled by "the most terrific scream that ever greeted my ears." As the noise seemed to come from the woods near his house, he proceeded to search for the source of the eerie screaming. He wrote later, "I immediately commenced searching round for the cause of this unearthly noise, but after a half hour's fruitless search, I resumed my labors ..."

The moment McHenri resumed his wood chopping, the ear-piercing shriek came again, and

THE REAL COWBOYS & ALIENS

looking up, he noticed a small humanoid about 50 yards away, standing between two trees. As he advanced toward what he referred to as the "miniature human," the creature fled. McHenry took off running and had to go "some distance" before overtaking the creature and capturing it.

McHenri described it as being eighteen inches in height, of the male sex, and with limbs that were in perfect proportion to its body. Its entire body, except for its face, hands, and feet, were covered with jet black hair. The description makes the creature sound almost like a miniature Bigfoot!

After capturing it, McHenri said, "The little fellow turned a most imploring look upon me, and then uttered a very sharp, shrill shriek, resembling the whistle of an engine."

Carrying it to his house, McHenri attempted to feed the creature some meat, but the humanoid refused to eat it. He then offered it water, of which it did drink a small quantity. Later he tried giving it "dried beach nuts," which the creature cracked and ate readily.

The newspaper account closes with McHenri inviting visitors to go see the creature on exhibit at his house. "Whoever may wish to see this strange specimen of human nature, can gratify their curiosity by calling at my house, in the eastern part of Waldoboro, near the Trowbridge tavern."

Then he added, "I give these facts to the public to see if there is anyone who can account for this wonderful phenomenon." The newspaper editor commented, "There are several persons who can

EARLY AMERICAN UFOS: 1800-1864

vouch for the truth of the statements [given by McHenri]."

Photo of Waldoboro (Circa 1869-1880) Taken by Asa H. Lane (Courtesy N.Y. Public Library)

Although the episode does not include a UFO sighting at all, the authors include it here because it occurred in an area that is historically a UFO hotspot and because the small humanoid described seems similar to creatures associated with UFO sightings. The being appears to have intelligence, despite being labelled a "wild man" by the media. It displays intelligence in discriminating between the types of food it ingests. Although it

THE REAL COWBOYS & ALIENS

does not speak a human language, it seems able to communicate by a series of shrieks and other vocalizations. Its height, 18 inches, is about half the size of most UFO occupants that have been observed over the years, which are typically three to four feet in height.

Aliens with hair or fur covering their bodies are rare among UFO reports, but interestingly, an incident involving creatures of this sort occurred in South America many years later, on November 28, 1954. On a rural road near Caracas, Venezuela, around midnight, two men, Jose Ponce and Gustavo Gonzalez, were driving along in a van when they were blinded by a bright light. They soon observed a relatively small ball of light, about three feet in circumference, bobbing in the air like a ship on the ocean waves.

Ponce stopped the van and Gonzalez exited the vehicle to examine the floating orb. To his shock he was knocked down by an unseen force -- no, not from the UFO, but from its occupant. Blinded and distracted by the orb's light, Gonzalez was unaware of the short, three-foot-tall creature that rushed him and pushed him over. Gonzalez looked at his attacker, which had the general appearance of a man, except that it was so short and covered in hair! Gonzalez tried to plunge his knife into the creature only to have it bounce off with no effect. Then a second "hairy dwarf" rushed Gonzalez, using something akin to a flashlight to blind and disorient him. Ponce got out of the van to aid his friend as two more dwarfs came out of the dark, this pair armed with rocks. Upon the sight of Ponce, the

EARLY AMERICAN UFOS: 1800-1864

little beings all raced back towards the orb of light and somehow jumped inside of it, despite the fact that it was in no way large enough to hold all four. To fans of science fiction, this brings to mind the fictional time machine TARDIS from the *Doctor Who* series. Perhaps the orb was a gateway into another dimension of space or time?

This wasn't the last that Venezuela would see of these strange creatures, however. The same craft and four beings attacked two hunters out in the woods several days later on December 12, attempting to abduct the men. Again, the odd beings were near indestructible. When one of the hunters whacked a creature with his shotgun, the weapon broke apart!

Six nights later, Jesus Paz exited a friend's vehicle to relieve his bladder in the bushes. His friends heard Jesus scream. They ran to the spot and found the man covered in gashes as though he had been attacked. Then a small, hairy man rushed out of the bushes and hopped into another small UFO which soared away into the night sky. This was the last known attack, and the creatures haven't been seen since.

Perhaps the tiny creature seen in Waldoboro, Maine, in January of 1855 was a smaller (younger?) version of this same species of alien? Again, it's interesting to note that the incidents in South America happened almost exactly 100 years after the one in Maine.

Whatever the Maine being was, unfortunately there is no further mention in the historical record

THE REAL COWBOYS & ALIENS

about this mysterious creature nor the person who found it, J. W. McHenri.

Efforts to find data on a person named "McHenri" were unsuccessful; however, there were a number of people with the surname "McHenry" living in Maine at the time of this incident. Originating in Ireland in the 1300s, the surname "McHenri" is an early form of the more common "McHenry." There are still instances today of the surname "McHenri" in the U.S.

Could the story of the "miniature human" have been a hoax intended to draw visitors to the vicinity of the "Trowbridge tavern," where they could go to parch their thirst or drown their sorrows after a fruitless search for the residence of Mr. J. W. McHenri, which may have been nonexistent? We suppose that is possible, although many of us probably "want to believe" that some time we might suddenly encounter an 18-inch tall humanoid skulking in a forest near where we live. And who knows, maybe this story really happened after all?

HUGE UFO IN OHIO
1858, Jay, Ohio

THE REPORTED SIGHTING of an immense aerial vessel over the town of Jay, Ohio, in the year 1858, continues to be a major puzzle to UFO researchers of today. The will to believe is strong on this one, and yet its authenticity has never been firmly established. The incident has been closely studied by many highly acclaimed researchers, including Jacques Vallee and Chris Aubeck, in their 2009 book *Wonders in the Sky: Unidentified Aerial Objects from Antiquity to Modern Times*. Although the origin of the tale is somewhat sketchy, the story just doesn't seem to want to go away. Although it has not been proven, neither have skeptics been able to discredit it.

THE REAL COWBOYS & ALIENS

The story goes that sometime around 1858, in a small Ohio town that no longer exists called Jay, several witnesses including Mr. Henry Wallace were startled by a large shadow and looked up to see an amazing sight. Above them was "a large and curiously constructed vessel, not over one hundred yards from the earth."

Similar Airship Seen in California Years Later

The principal eyewitness, Wallace, later stated that he believed the vessel to be from Venus, Mercury, or the planet Mars. He described it as follows: "The vessel was evidently worked by wheels and other mechanical appendages, all of which worked with a precision and a degree of beauty never yet attained by any mechanical skill upon this planet ... This was no phantom that disappeared in a twinkling ... but this aerial ship was guided, propelled and steered through the atmosphere with the most scientific system and

EARLY AMERICAN UFOS: 1800-1864

regularity, about six miles an hour, though, doubtless, from the appearance of her machinery, she was capable of going thousands of miles an hour."

As the strange aerial vessel passed above their heads, the witnesses noticed several "tall people" that were aboard the ship. Wallace said these strange individuals were possibly "on a visit of pleasure or exploration, or some other cause."

The story's level of detail and seeming authenticity has had large appeal for Ufologists over the years. However, its source remains to this day, highly suspect. It was first published in 1858 in a book containing medical advice (including lots of medical quackery) that fell more in the area of the occult than on the scientific. The book by Dr. William Earl was called *The Illustrated Silent Friend, embracing subjects never before scientifically discussed.*

On the web site *Rense.com* in 1999, author Jesse Glass, Jr., published a detailed account of this sighting in his article, "A UFO Visits Ohio in 1858." In it, Glass describes the book by Dr. William Earl, in which the sighting first appeared, as follows: "... Dr. Earl delivers a breathtaking range of quaint and curious lore. Sandwiched in between advertisements for male safes [condoms] made of white Indian rubber, and herbal cures for gonorrhea, were recipes for making wood more durable than iron and practical advice for would-be, part-time mesmerists.... Among all the snake-oil claptrap contained in this entertaining bit of Americana is one entry that, in its strangeness, and

THE REAL COWBOYS & ALIENS

its use of specific names and places, stands out from the rest. On pages 253-256, we find this startling entry:

Do the Inhabitants of other Planets ever Visit this Earth? I propose in this connection to make a few remarks on the following: Mr. Henry Wallace and other persons of Jay, Ohio, have recently detailed to me the annexed. There are thousands of such cases on record. These gentlemen state, that sometime since on a clear and bright day, a shadow was thrown over the place where they were; this necessarily attracted their attention to the Heavens, where they one and all beheld a large and curiously constructed vessel, not over one hundred yards from the earth. They could plainly discern a large number of people on board of her, whose average height appeared to be about twelve feet. The vessel was evidently worked by wheels and other mechanical appendages, all of which worked with a precision and a degree of beauty never yet attained by any mechanical skill upon this planet.

Now, I know that thousands will, at this recital, cry humbug, nonsense, lunacy, etc., but I know that there are other thousands who will read and reflect. It is for these latter thousands that I write. Once upon a time there appeared a celebrated reformer, who arose among the people and taught a new doctrine, that from its reasonableness and its simplicity, electrified the hearts of the thinking people. But the party who didn't think, and who hated reason, and new ideas, cried out, away with him to the crucifixion. And they did crucify his

EARLY AMERICAN UFOS: 1800-1864

body, but they have not yet succeeded in crucifying the reason, and new facts and ideas that he taught.

Airship Illustration (Public Domain Photo)

In view, then, of the above, I venture to advance the following remarks: I believe that the time will come when all of the inhabitants of all worlds or planets in the solar system, will regularly visit each other when in the fullness or fruition of things, an interchange of ideas and commodities, visiting and greetings between the respective inhabitants of all worlds or planets, will be common and universal. I believe that the grand aspirations of an advanced humanity on this earth, is not without a good cause and a good reason.

I believe that when the respective atmospheres seen surrounding the different planets in the solar system, indeed, of every part of the universe, shall

THE REAL COWBOYS & ALIENS

have passed into the highest condition of excellence and purity of which it is capable, that it will then give life to a more exalted and finished condition of genera and species, or inhabitants. That all of the planets are now inhabited by a kind of being suited to their respective planetary and electrical conditions, is, I think, certain. And that the inhabitants of thousands of these worlds, that roll with eternal beauty throughout the boundless regions of the immensity of space, have attained that advanced condition in their planetary being, I have no doubt, whatever.

And that this ship which Mr. Wallace and others saw, was a vessel from Venus, Mercury, or the planet Mars, on a visit of pleasure or exploration, or some other cause; I myself, with the evidence at hand, that I can bring to bear on it, have no more doubt of, than I have of the fact of my own existence. This, mind, was no phantom that disappeared in a twinkling, as all phantoms do disappear, but this aerial ship was guided, propelled and steered through the atmosphere with the most scientific system and regularity, at about six miles an hour, though, doubtless, from the appearance of her machinery, she was capable of going thousands of miles an hour, and who knows but ten thousand miles an hour. What can be more wonderful as an illustration, than the Electric Telegraph to connect the old world with the new. And why then, may not the scientific geniuses of other planets have done as much as ours have?

EARLY AMERICAN UFOS: 1800-1864

Besides this, if I had room, I could draw an argument from the electrical condition of the media existing between the planets, to show that a body once in motion at a given distance from a planetary body in space, will move with nearly the speed of electricity till it meets again the resisting media, or atmosphere of another planet or body in space. That all of this knowledge, and a million times more, may be known to some of the exalted beings of other planets in space, I have no doubt. But as I was saying, this aerial ship moved directly off from the earth, and remained in sight, till by distance she was lost to the view. The foregoing is my firm and decided conclusion and belief in this matter."[10]

Over the years since Glass wrote this article, attempts to verify the validity of the information contained in this account have met with mixed success. Historical records have proven that a post office existed in the town of Jay, Ohio, from 1839 to 1842. The town's status in 1858 is less certain. Also, census records show a number of individuals named Henry Wallace did live in the area at the time, although none of the persons seems to exactly match the eyewitness of this UFO sighting.

Since the original source of this report is dubious and attempts to verify its details have failed, the sighing remains firmly in the "unverified" category. And yet, it continues to nag at us for attention and

---

[10] https://rense.com/ufo3/visits.htm

has been described by some as one of the most interesting UFO cases of the nineteenth century.

NIGHT OF THE UFOS
July 20, 1860
Northeastern United States

IN THE EARLY TO MIDDLE 1800s, there were only four possible explanations for illuminated objects moving across the night sky: (1) comets, (2) untethered hot air balloons carrying very bright lanterns, (3) brightly lit birds, or (4) something unexplained and possibly extraterrestrial. Okay, maybe hot air balloons and birds can be eliminated from this list! With this in mind, we focus on a singularly fascinating case that some people refer to as a meteor event, but which many others are not sure about.

On the night of July 20, 1860, in the northeastern United States, a swarm of luminous orbs moved across the sky slowly, west to east, across hundreds of miles, from northern Michigan to Long Island, New York, and then on to the Atlantic Ocean. They were seen and marveled at by thousands of people. The two main orbs were described as nearly the size and brightness of the moon, and many observers saw numerous smaller orbs, perhaps as many as fifty, following the larger ones. The swarm illuminated cities and towns in an eerie bluish white light and terrified spectators who feared Doomsday had arrived. The awe-inspiring display made national headlines for weeks and is believed to have inspired the Walt Whitman poem "Year of the Meteors." It is even memorialized in a painting by American landscape painter Frederic Church. Since no objects struck the ground and no fragments were ever recovered, the event was considered an "unexplained atmospheric phenomenon" for 150 years, until in the year 2010, scientists guessed that the event was probably an extremely rare "Earth-grazing meteor procession."

Despite this latest attempt to scientifically explain the mystery 150 years after the fact, some people still believe that there was something strikingly unearthly about the swarm of orbs seen by thousands in some of America's most populous cities back in 1860. Instead of a rare meteor event of the kind that scientists think has only occurred twice before in human history, perhaps this was something else? What if instead of meteors, what people saw was a fleet of UFOs streaking across the lower atmosphere before disappearing back up into the heavens?

EARLY AMERICAN UFOS: 1800-1864

Painting of the Phenomenon by Frederic Church

On July 21, 1860, the day after the phenomenon, articles about it appeared in almost every newspaper in the northeastern U.S. with many eyewitness accounts. The stories of amazed people who witnessed the spectacle continued for weeks. In the *Brooklyn (N.Y.) Daily Eagle*, eyewitness Thomas Prosser of Long Island wrote, "It came from the northwest ... when it appeared like two immense rockets, accompanied by a few brilliant scintillations. It disappeared in the southeast."

Another New York witness said, "... two balls of fire arose in the west, moving slowly southeast, when seemingly over the East River, the balls exploded and shot with lightning speed across the heavens. The sight was a grand one, two large meteors flew through the air, lighting up the city as if each had been a moon with 50 little red-faced meteors chasing after the greater splendor" (7/21/1860).

At Danville, Pennsylvania, an eyewitness wrote that the phenomenon emitted "as much light as full moon." The witness also noted a sonic boom, saying "Some minutes after it disappeared, a sound resembling thunder was distinctly heard." (7/21/1860).

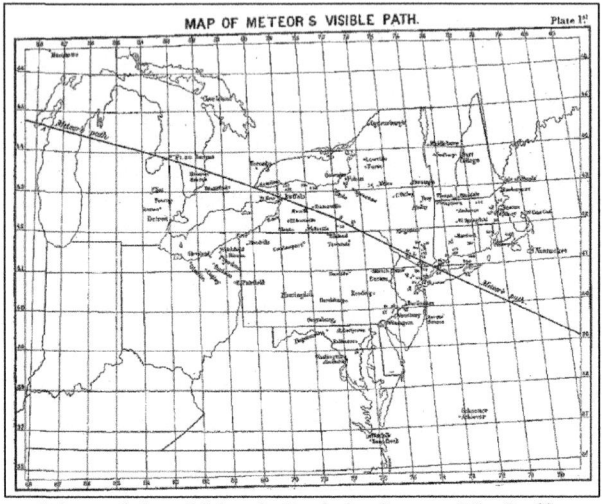

Map of the Object's Path by James H. Coffin (1806-1873)

A witness in Baltimore, Maryland commented that the object was moving "with a motion entirely different from any meteor I have ever seen mentioned anywhere." His description was impressive: "Its first appearance was in the shape of a ball of greenish light, half the diameter of the moon, around which was a large halo ... At the distance of a mile or a mile-and-a-half from the city, it looked like a long, ragged tongue of bluish flame, sailing slowly and majestically along."

An observer in Newark, New Jersey, said, "It appeared composed of two distinct bodies of equal

EARLY AMERICAN UFOS: 1800-1864

size, and surmounted by a very brilliant blue and yellow light ... it might have been mistaken for an immense rocket, both from its appearance and the remarkable light.... The meteor appeared so altogether different from all meteors noticed before, that we are at a loss what to say about it ...." (7/23/1860).

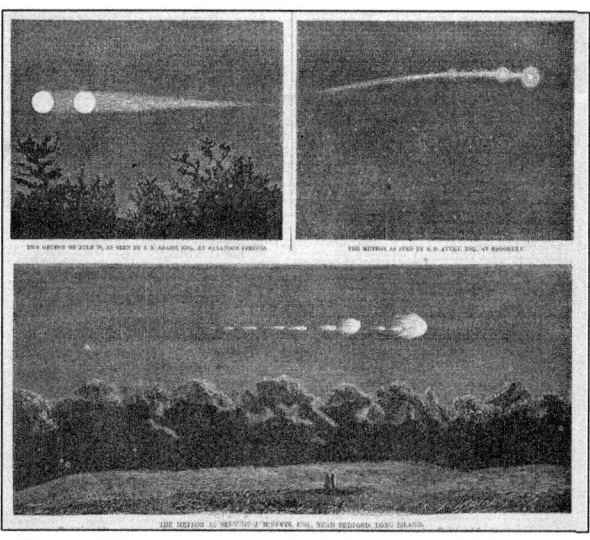

From "Harper's Weekly," August 4, 1860, p. 1.

In *The Daily Evening Express* of Lancaster, Pennsylvania (7/21/1860), a reporter is quoted as saying that the swarm "contained two or three balls of a beautiful bluish light." The newspaper notes that many other observers in Lancaster also commented on the blue light, which was said to have illuminated the streets and buildings down below the objects. Another Lancaster witness said, "When first seen, it displayed a tail that appeared

THE REAL COWBOYS & ALIENS

to the naked eye to be fifty or sixty feet in length, and in its course, the tail decreased, leaving sparks and streaks of fire in its rear." Many witnesses throughout the Northeast used the phrase "like a rocket" to describe what they saw. "It appeared to be three quarters of a mile high and at first had the appearance of a large fish."

> METEORIC PHENOMENON.—Shortly before 10 o'clock last evening a meteor of great size and brilliancy passed over this City, taking a direction from west to east. It was first observed, so far as we could learn, by passengers on the Jersey City ferry-boats, to whom it appeared to have shot out from behind the houses in something of a circular course. It seemed to pass directly over the City Hall Park, when it formed into two distinct heads, one shooting ahead of the other, and leaving a long train of fiery rain between them, similar to the appearance of a rocket after it has burst. The light was of a dazzling blue and white color, and during its flight it illuminated every thing in its train.

The New York Times -- 21 Jul 1860, Page 8

Regarding its elevation, most witnesses felt it was only a few hundred feet above them. One witness stated that if a five-story building had been in its wake, the building would have been demolished. Others put the elevation at between 200 feet and one mile, which does not seem to support the theory that it was a meteor that grazed the Earth's atmosphere before skipping off again into space.

Observing the phenomenon, an editor for the *Philadelphia Ledger* wrote, "... an object about the size of the full moon, and as bright ... traversed in a direct easterly line the whole extent of visible space, dropped fire apparently in its course, like a

EARLY AMERICAN UFOS: 1800-1864

rocket, till it passed so far eastward as to resemble a red ball, about twice the size of the planet Mars."

In a report appearing in the *New York Times* on July 21, the following was stated: "Shortly before 10 o'clock last evening (July 20), a meteor of great size and brilliancy passed over this City, taking a direction from west to east ... It seemed to pass directly over the City Hall Park, when it formed into two distinct heads, one shooting ahead of the other and leaving behind a long train of fiery rain between them, similar to the appearance of a rocket after it has burst. The light was of a dazzling blue and white color, and during its flight, it illuminated everything in its train."

The effect of the phenomenon upon the spectators watching below was captured by a witness in the July 23rd issue of the *Baltimore Sun*: "During the time of its nearest approach to the earth, a ghastly phosphorescent [blue] light was reflected against prominent buildings so distinctly as instantly to attract the attention of every one outdoors; and the whole singular appearance drew loud exclamations of surprise and wonder, and, in some cases, of real terror, from everyone who beheld it."

The New York Post of July 23 reported: "Its brilliancy was so great that people at once supposed there was a fire nearby, but looking up, they saw two balls of flame coursing across the sky from the northwest, and going toward the southeast ... When first seen, it appeared like a blue star surrounded by a thick mist ... which on nearing the zenith changed to a red ball of flame; this soon split, the

THE REAL COWBOYS & ALIENS

two balls keeping near together until lost in the distance."

On August 2, in a letter to the *New York Tribune*, renowned Harvard astronomer George Phillips Bond, who in 1850 had taken the first photo ever of a star (Vega), summarized the path of the 1860 anomaly, based on eyewitness accounts. The swarm of objects was first spotted in the vicinity of the Great Lakes near northern Michigan and progressed to the east-southeast. Travelling in a straight line, it passed over Lake Huron, southwestern Ontario, Lake Erie, southwestern New York state, the northeastern part of Pennsylvania, southeastern New York, the southwest corner of Connecticut, Long Island Sound, and Long Island. Continuing to the Atlantic Ocean, it was last seen 300 to 400 miles out to sea.

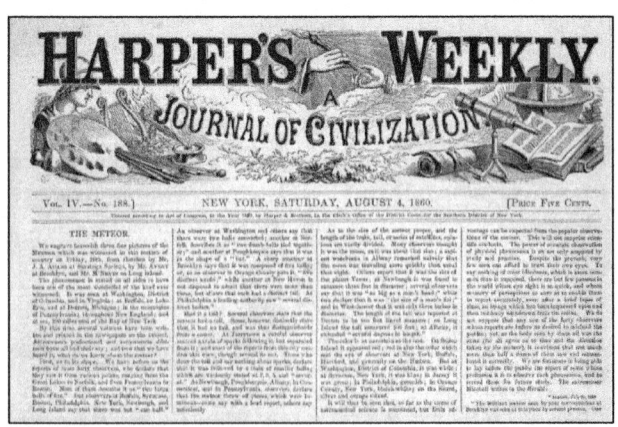

From "Harper's Weekly," August 4, 1860, p. 1.

Interestingly, a number of similar "meteor" sightings were made on August 2, 1860, several

EARLY AMERICAN UFOS: 1800-1864

hundred miles south of the July 20th path, in both Virginia and Georgia. In 1964, UFO researcher Richard Hall of the National Investigations Committee on Aerial Phenomena (NICAP) classified the August 2nd sighting in Norfolk, Virginia, as possibly being multiple UFOs. A report in the *Norfolk Herald* said, "The meteor, or rather meteors – for, like that which was seen a few nights previously [July 20], it represented a duplicate appearance – were each about the size of a butter keg, and not unlike that object in form, though it slightly rounded at the ends. Starting into view at a point about W.N.W., and taking a northerly direction, they sped rapidly with an undulating motion, rising and subsiding twice so as to describe in their course a double arch of easy and graceful cure, preserving their brilliance to the end, and finally disappearing at several degrees above the horizon."

"One of the lights was of a clear red and the other a greenish complexion; and both, as they coursed along, emitted resplendent flashes of the same beautiful hues, while their track was marked by a sparkling train of light similar to that left by a large and brilliant rocket." Richard Hall of NICAP was impressed that the two objects moved together and that they exhibited an "undulating motion." The chances that this event was another "meteoric progression," just days after the July 20th event seem very remote.

150 years after the July 1860 amazing aerial display, a team of scientists at Texas State University in San Marcos, Texas, determined that

what people saw in 1860 was the combination of two extremely rare celestial events -- an "Earth-grazing meteor," which strikes the atmosphere but sweeps back into space -- and a "meteoric procession," which occurs when a meteor breaks up in the atmosphere, creating multiple meteors traveling together in a nearly identical path. In publishing their research in *Sky & Telescope* (July 2010) magazine, the scientists admitted that meteoric processions are extremely rare and only two others seem to have ever occurred in human history, one in 1783 and another in 1913. No "Earth-grazing meteoric procession" had been previously heard of, prior to their findings.

In conclusion, the 1860 event was clearly either an extremely rare, once-in-a-millennium "Earth-grazing meteoric procession," or it was something totally unexplained. After reading all the eyewitness accounts provided herein, our esteemed readers are left to decide which scenario seems to make more sense.

UFOs OVER THE MISSOURI RIVER
September 24, 1860
Nebraska City, Nebraska

MANY OF THE unidentified aerial phenomena seen in the early 1800s consisted of just a single object, usually described as looking sort of, but not exactly, like a comet, meteor, or other astral object. However, there was a singularly unique sighting of multiple objects all flying together in formation, on September 24, 1860 in Nebraska City, Nebraska.

The primary witness was Joel Draper, a gardener living in Nebraska City, and the objects were also seen by a ferry boat operator identified as "Mr. Beebout." The incident began ten or fifteen minutes before sunset as Draper boarded a ferry boat at the Missouri River.

THE REAL COWBOYS & ALIENS

Missouri River in Nebraska (Courtesy Scott Redd)
[CC BY-SA 2.5, https://commons.wikimedia.org/w/index.php?curid=2614773]

Suddenly, the ferryman, Mr. Beebout, noticed something in the sky and called it to the attention of Draper, who later wrote, "As I came on to the ferry boat on Monday, about sunset, from the east bank of the river, Mr. Beebout, the ferryman, directed my attention to a bright spot in the horizon near the point where the sun had been, when one hour high."

Both men gazed in amazement at the bright object, whose "size, color, brightness and shape resembled one-fourth of the sun taken from its edge...."

They noticed a second, smaller object to the right of the first and a bit higher. And they also observed a third, even smaller object slightly above the first. The relative sizes of the objects could have been deceptive to viewers on the ground, depending on the actual distance from the viewer to each of the three objects. It is possible that all three objects

EARLY AMERICAN UFOS: 1800-1864

were actually identical in size but were flying at three distinctly different distances from the observers down below.

Draper noted that the objects were the same size, color, and brightness of the sun, and their shape looked like a "wedge" amounting to one-fourth of the sun's disc as viewed by a ground observer. Therefore, the objects were pie wedge-shaped or possibly even triangle shaped.

All three objects were moving toward the south with tremendous speed, relative to the position of the sun. Draper said, "All of these we discovered to be moving towards the south, or to the left of their former position, with great rapidity, as if they were as far off as the sun."

Interestingly, the witness said that as the three objects moved "they all retained the same relation to each other as when they first appeared." It seems these constituted a formation of UFOs, moving across the sky together, all at different altitudes relative to the ground. Alternately, the three objects may have been part of a much larger object that was

THE REAL COWBOYS & ALIENS

not visible to the observers because it was dark and blended in with the dark sky. In more recent decades, UFOs have been seen that appear to be a formation of small bright spheres but turn out to actually be lights that are on the hull of a much larger object, as in the image shown below.

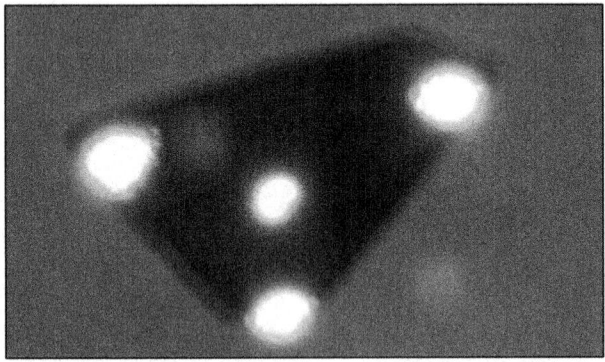

A Dark Triangle UFO Seen in Belgium in 1990

Also of interest, Draper observed that the light emanating from the objects was so great that it actually illuminated the surrounding area, as if it were sunlight. Draper said, "This took place after sunset, but by means of the brightness of these bodies, it was as light as some ten or fifteen minutes before sunset." These were, by no means, dim lights.

Eventually, the three objects "disappeared from our view, by the intervening bluffs at the levee." In concluding his account of the sighting, Draper added, "They could not have been sun dogs or mock suns, for they remain, as long as, they continue, in the same relative position to the sun."

EARLY AMERICAN UFOS: 1800-1864

Singular Celestial Phenomenon.—A remarkable and interesting phenomenon was observed by a number of people at Nebraska City, K. T., on Monday evening, Sep. 24, about sundown or shortly after. Joel Draper, who witnessed it, furnishes the following account of the appearance presented:

As I came on to the ferry boat on Monday, about sunset, from the east bank of the river, Mr. Beebout, the ferryman, directed my attention to a bright spot in the horizon near the point where the sun had been, when one hour high. While gazing with amazement at *that*, which in size, color, brightness and shape resembled one-fourth of the sun taken from its edge, we soon discovered another spot further to the right and a little higher, which was about one-third of the size of the first; then another directly above the first, one-third the size of the second. All of these we soon discovered to be moving towards the south, or to the left of their former position, with great rapidity, if they were as far off as the sun. They soon disappeared from our view, by the intervening bluffs at the levee. As they moved, they all retained the same relation to each other as when they first appeared.

This took place after sunset, but by means of the brightness of these bodies, it was as light as some ten or fifteen minutes before sunset. There was no clouds or vapors in the sky in that direction. They could not have been sun dogs or mock suns, for they remain, as long as they continue, in the same relative position to the sun.

Will not some one of your readers who witnessed this remarkable phenomenon till it disappeared from the heavens give us a more complete description of it, and also his views of the cause?

JOEL DRAPER.

NEBRASKA CITY; Sep. 26, 1860.

The Brooklyn (N.Y.) Daily Eagle - Oct 12, 1860, p. 2

THE REAL COWBOYS & ALIENS

Then, the article closes with Draper imploring the readers of the newspaper, "Will not someone of your readers who witnessed this remarkable phenomenon till it disappeared from the heavens give us a more complete description of it, and also his views of the cause?"

The site of this UFO incident – Nebraska City – is located on the extreme eastern edge of Nebraska in Otoe County and is situated on the western bank of the Missouri River. The Lewis and Clark Expedition, traveling west along the river, came across the area in 1804, finding a number of Native Americans in the area. The city was also the site of the U.S. Army fort named Old Fort Kearney, built in 1846 and used for several years before being abandoned. Nebraska City became the first incorporated city in the state of Nebraska by a special act of the Nebraska Territorial Legislature in 1855.

Aside from being the year of the UFO sighting, 1860 is also known for a great tragedy that occurred in Nebraska City. A huge fire broke out on May 12 of that year, destroying almost the entire business district of the city. Among the buildings that were consumed was the post office, the government land office, and the Nuckolls Hotel. The losses were estimated to be $120,000, which equals 3 and a half million in 2015 U.S. dollars.

The 8th U.S. Census (1860) found a total of 1,922 persons residing in Nebraska City. Among the residents was a man named Joel Draper, listed as a 46-year-old white male residing in the "3rd Ward, Otoe, Nebraska Territory." He was born in

EARLY AMERICAN UFOS: 1800-1864

Massachusetts in approximately 1814 and listed his occupation as a gardener. His wife is listed as Sarah, and a 14-year-old daughter, Flora, is also listed. The value of Draper's dwelling is given as $6,600 and his personal estate is $100.

Excerpt from 1860 U.S. Census

The story of the Nebraska City sighting first appeared in the October 12, 1860 issue of *The Brooklyn (N.Y.) Daily Eagle* newspaper. The case was also listed among the most interesting UFO cases of the nineteenth century in the 2009 book *Wonders in the Sky: Unidentified Aerial Objects from Antiquity to Modern Times* by Jacques Vallee and Chris Aubeck. This could be one of the most significant UFO sightings of its time period, and it contains many elements that would not be commonly seen in UFOs until 100 years later, such as moving in formation at great speeds and having a shape that was not reported in UFOs until late in the twentieth century.

This case was one of two fascinating sightings of multiple UFOs that were reported in the year 1860. The second case, in Wilmington, North Carolina, two months later, also featured three luminous

THE REAL COWBOYS & ALIENS

objects seen by several witnesses, as we will learn in the next chapter. During this part of America's history, a single sighting of more than one unidentified aerial phenomenon was extremely rare.

22
SPACESHIPS OVER THE OLD SOUTH
November 10, 1860
Wilmington, North Carolina

ANOTHER SIGHTING of multiple UFOs occurred in Wilmington, North Carolina, on Saturday, November 10, 1860. The location of this event is sometimes erroneously given as Washington, D.C. or Washington, North Carolina, but it actually happened in Wilmington, which is a port city in the southeastern portion of the state. It is the county seat of New Hanover County.

This UFO sighting was reported in the *Wilmington (N.C.) Herald*, *The Times-Picayune (New Orleans, LA)*, *The Newbern (N.C.) Weekly Progress*, and *The Brooklyn (N.Y.) Daily Eagle*. The original report states that a group of witnesses, including the reporter who filed the story, noticed

THE REAL COWBOYS & ALIENS

a peculiar object in the sky at about sunset. From the ground, it looked like either a balloon [orb] or a piece of chalk [cigar shape or rod]. The shape of the object probably changed slightly as it moved and adjusted its angle, going from orb to elongated orb or cigar shape.

Sketch of the City, circa 1800s
(N.C. State Archives)

The UFO was extremely bright and highly visible in the daytime sky. "Notwithstanding [that] the light of day was still strong and clear, the illumination of the object was brilliant and distinct...." The object's shininess and high visibility suggests that its hull may have had a shiny, metallic finish.

At first, the strange object was observed moving very rapidly to the southwest. Its path was roughly east to west. But suddenly, during his observation of the object, the newspaper reporter noticed that the object changed its course from southwest to due west. This change of direction would favor the theory that this object was under intelligent control.

Some of the observers stated to the reporter that two more identical objects had already passed by overhead before the reporter arrived. So, a total of

EARLY AMERICAN UFOS: 1800-1864

three objects were seen flying together from east to west, in a sighting that has similarities to what was seen by gardener Joel Draper two months earlier in Nebraska City, Nebraska, which is over 1,300 miles to the northwest. The objects Draper saw were headed roughly to the east.

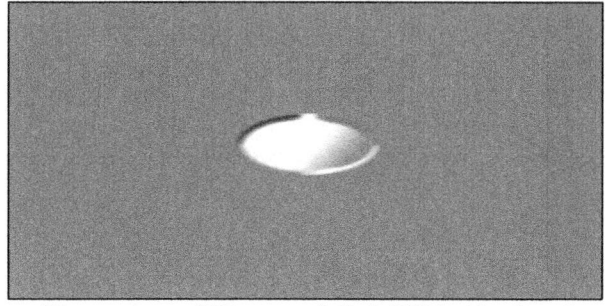

Illustration of Elongated Orb UFO (Authors)

This is another sighting that is significant because it is quite unlike most of the other North American sightings of the nineteenth century. The object is not described as being an airship, comet, fireball or meteor. It is distinctly an orb or elongated orb [saucer] that, because of its brightness, probably had a shiny metal exterior. The sighting is also significant because three objects were seen at the same time by multiple witnesses.

The newspaper reporter added a kind of amusing comment at the end of the article, stating, "It was in all probability a *parhelion*, or mock-sun, which is the original and highest species of the genus 'toady [a lackey or minion],' inasmuch as it only follows in the wake of and tries to look and act like the god of the day [Sun]. *Wikipedia* notes, "A

THE REAL COWBOYS & ALIENS

sun dog (or sundog) or mock sun, formally called a *parhelion* in meteorology, is an atmospheric optical phenomenon that consists of a bright spot to one or both sides of the Sun. Two sun dogs often flank the Sun within a 22° halo. The sun dog is a member of the family of halos, caused by the refraction of sunlight by ice crystals in the atmosphere. "

> **SIGNS IN THE HEAVENS.**
>
> A singular phenomenon was witnessed by some of our citizens on Saturday evening last, just about sunset. An object about the size of a balloon (or peice of chalk,) and very much of the appearance of a balloon, was seen moving with great rapidity in a Southwesterly direction; and, notwithstanding the light of day was still light and clear, the illumination of the object was brilliant and distinct as a balloon at night. We heard, while gazing at this wonder, that two other similar ones had passed previously. The one we saw, after moving Southwesterley, at an angle with the path of the sun, took a course directly West, and straight from us; fading gradually, and very rappidly, until lost from sight. It was in all probability a *parhelion*, or mock-sun, which is the original, and highest species of the genus "toady," inasmuch as it only follows in the wake of and tries to look, and act like the god of day.— *Wil. Herald.*

Newbern (N.C.) Weekly Progress,
11-20-1860, p. 4

The reporter's comment about a "toady" indicates that he understood the object he saw was not typical of sun dogs, as this object was not

EARLY AMERICAN UFOS: 1800-1864

stationary at one side or the other of the sun. The object he saw was moving independently of the sun although appearing to follow in its wake, thus he describes it as being a "toady" of the sun.

*Activity at the Port in Wilmington,
North Carolina, circa 1900*

The story received widespread attention from the media of the time. The same article from the *Wilmington Herald* appeared in several other newspapers, some as far away as New Orleans, Louisiana:

THE REAL COWBOYS & ALIENS

> *Signs in the Heavens.*—The Wilmington (N. C.) Herald, of the 12th, says:
>
> A singular phenomenon was witnessed by some of our citizens on Saturday evening last, just about sunset. An object about the size of a balloon (or piece of chalk), and very much of the appearance of a balloon, was seen moving with great rapidity in a southwesterly direction; and, notwithstanding the light of day was still strong and clear, the illumination of the object was brilliant and distinct as a balloon at night. We heard, while gazing at this wonder, that two other similar ones had passed previously. The one we saw, after moving southwesterly, at an angle with the path of the sun, took a course directly west, and straight from us; fading gradually, and very rapidly, until lost from sight. It was, in all probability, a parhelion, or mock-sun, which is the original, and highest species of the genus "toady," inasmuch as it only follows in the wake of and tries to look, and act like the god of day.

The Times-Picayune (New Orleans, LA), 11-18-1860, p. 9

A Train Leaves Wilmington, N.C. in 1909

23
MYSTERIOUS AIRSHIP OVER NEW YORK

October 5, 1861
New York, New York

WITH THE ONSET of the U.S. Civil War in 1861, the nerves of the common people in New York City were a bit frayed, as they heard of the activities of the Confederate Army just to the south. An article in the *New York Times* on October 4, 1861 stated that a Confederate tethered hot-air balloon had been spotted in the direction of Springfield [probably in Pennsylvania] but that "at so great a distance its occupants could hardly have obtained any information as to our forces or movements."

Interestingly though, on the following day, the *Times* reported that possibly a Confederate hot air balloon had become untethered and had flown

THE REAL COWBOYS & ALIENS

directly over New York with two men in it, which some UFO researchers have said seems to be unlikely since many precautions were taken to prevent the loss of these types of balloons.

> **A REBEL BALLOON.**
> The rebels have a balloon, which was seen this P. M. in the direction of Springfield, but at so great a distance that its occupants could hardly have obtained any information as to our forces or movements.

> **A BALLOON ASTRAY.**
> A mysterious balloon, with two men in it, passed over this city about 5 o'clock this afternoon, from west to east, at a great height. The sensationists have it that it was in the service of the traitors, and that it had C. S. A. conspicuously painted on its sides; others say, which is more probable, that it was one of Prof. LOWE's, which had parted its fastenings and started off on its own hook.

The New York Times, 10-5-1861, p. 1

Many researchers have classified this as most likely a legitimate UFO sighting due to the fact that untethered balloon flight was still highly experimental, hence the title of the article, "A Balloon Astray." The writer states that it could be a balloon that "parted its fastenings" and headed off on an unplanned trajectory through the air. Another argument against it having been a hot air balloon is that the object, which was moving west to east across the skies of New York, was said to be traveling "at a great height."

EARLY AMERICAN UFOS: 1800-1864

Professor Thaddeus Lowe Standing to the Right of a Balloon as it is Being Inflated near Gaines Mill, Virginia, May 1862.

The Confederate States were known to employ tethered balloons to spy on enemy activities and troop movements. The North also had a balloon program under the direction of Professor Thaddeus Lowe, who was appointed Chief Aeronaut of the Union Army Balloon Corps by President Abraham Lincoln in July 1861. The article about the mysterious balloon speculates that "others say, which is more probable, that it was one

THE REAL COWBOYS & ALIENS

of Prof. Lowe's [balloons], which had parted its fastenings and started off on its own hook."

Professor Thaddeus Lowe, 1865

But Lowe's balloons, like their Confederate counterparts, were extremely primitive affairs. They mostly carried one man aloft and did not normally drift untethered. Did these balloons ever break loose and escape? Surely they did, but mostly the balloons were used to ascend and descend. If they flew untethered, they could only go where the wind currents took them and landing back on the ground safely was often problematic.

EARLY AMERICAN UFOS: 1800-1864

So, if the "mysterious balloon" that flew over New York City on October 5, 1861 was not a Civil War era balloon, what could it have been? Is it possible that the object sighted was a total unknown and that the "men" in it were not entirely human? Well, probably not, as the Union Army revealed in the *New York Times* on the next day. They said it was the military balloon *Saratoga*, designed by freelance balloon inventor John La Mountain.

> **THE VAGRANT BALLOON.**
> It is said at the War Department that it was La Mountain's balloon, the *Saratoga*, which passed over Washington yesterday. Thus the reports and speculations that it was a vagrant rebel balloon are dissipated.

The historical record confirms that La Mountain's balloon was being flown experimentally during October 4th and 5th. So, it is certainly possible that what people saw over New York was simply La Mountain's ship *Saratoga*. Just

THE REAL COWBOYS & ALIENS

over a month later, the *Saratoga* was blown from its moorings and was lost behind enemy lines, thus ending La Mountain's efforts in the war.

In this case, the "mysterious balloon" seen over New York on October 5, 1861, was almost certainly La Mountain's *Saratoga*. However, some die-hard Ufologists remain convinced today that it was something much more otherworldly!

John La Mountain

HOUSTON, WE HAVE GIANT ALIENS!
September 1862, Indian Ocean

VETERAN RESEARCHER Kevin Randle, known for having been one of the first investigators of the 1947 UFO crash near Roswell, New Mexico, also looked into a very unusual reported UFO crash from 1862. In his book *When UFOs Fall from the Sky*, Randle briefly discusses this extremely odd case and concludes that it must have been a hoax. Still, it is worth our taking another look, especially since it is one of the most incredibly fascinating UFO stories, even if you agree with Randle that it must be completely fictional!

The story first appeared on Sunday, May 2, 1897 in the *Houston (Texas) Post* newspaper, although the incident occurred on a ship out at sea, in the Indian Ocean to be exact. Referring to an event that took place in 1862, the writer of the article was motivated to publish it in 1897 because of the many

THE REAL COWBOYS & ALIENS

reports of mysterious "airships" being seen all over the United States in 1897.

> **AIRSHIP OF THE PAS...**
>
> El Campo, Texas, April 3...
> weekly paper an opinion a...
> professor of astronomy that the...
> much has been published about...
> the planet Mars.
>
> There is an old sailor living ...
> Campo with his daughter who ...
> claimed that he had not only seen...
> vessel, but had actually seen people...
> another world. His immediate...
> have known of the circumstance for...
> years, but he says the story has never...
> published.
>
> The name of the old gentleman...
> Oleson and for many years he was a...
> swain in the Danish navy, but at the...
> he saw the airship he was mate...
> Danish brig Christine.
> In September, 1862, the Christine...
> wrecked in the Indian ocean on an...
> rock or island several miles in...
> This rock is set down in charts of...
> ocean, but is not mentioned in...
> raphies.

Portion of the Houston Post Article, May 2, 1897

The author is identified as John Leander, and he claims to have been told the story by "an old sailor" identified only as "Mr. Oleson," living with his daughter near El Campo, Texas, located 70 miles southwest of Houston. According to Leander, Mr. Oleson not only saw a strange airship with his own eyes, but he also saw the creatures from that ship, who were clearly from another world.

Realizing that many readers would not believe Oleson's story, the author writes, "... many believe those airship stories to be fake. That may be so, but the story now told for the first time is strictly true. While Mr. Oleson is an old man, he still possesses every faculty and has the highest respect for truth

EARLY AMERICAN UFOS: 1800-1864

and veracity." He adds that a number of the leading citizens heard the old sailor's story and saw evidence that it was true – "Quite a number of our best citizens, among them Mr. Henry Hahn, Dr. H. C. Carleton, Green Hill and Scott Porter...." Leander adds that Oleson had previously told the story to his "immediate relatives," but it had never been published.

```
GOODS ENTERED AT THE CUSTOM HOUSE
               YESTERDAY.
             IMPORTS FOREIGN.
Christine (Danish ship), Frost, from Tonning, 200 bags
  wheat flour, 500 cwts—Moxey and Winter.

            IMPORTS COASTWISE.
Eagle (s.s.), 215 tons, Deadrick, from Hull, sundry Bri-
  tish goods—Dundee, Perth, and London Shipping
  Company.
Hamburg (s.s.), 305 tons, Speedy, from London, 5 serons
  almonds, 5 hhds and 300 loaves sugar, 400 bags
  guano, 1700 bales jute, 2 hhds wine, 92 chests tea,
  1 tierce and 10 bags coffee, sundry other goods—
  Dundee, Perth, and London Shipping Company.
Rival, 113 tons, Smith, from Seaham, 180 tons coals—
  P. M. Duncan.
```

The Courier and Argus of Dundee of Tayside, Scotland (March 18, 1862, p. 4)

After serving as a boatswain in the Danish navy for many years, Oleson signed on to be a mate on a Danish brig named *Christine*. It was while serving on *Christine* that the encounter with the airship and its unearthly occupants happened. We found in the historical record that a Danish ship named *Christine* did in fact exist in 1862, according to *The Courier and Argus* of Dundee of Tayside, Scotland (March 18, 1862). The ship is listed as arriving at port in Tayside from Tonning, Germany, carrying

THE REAL COWBOYS & ALIENS

200 pounds of wheat flour and captained by "Frost."

In September 1862, *Christine*, while sailing in the Indian Ocean, encountered a furious storm that raged for hours, causing the ship to be swept far from her course. As Oleson and the other sailors fought to maintain control of their ship, an immense rock suddenly loomed directly ahead of the vessel amid the tumultuous waves. The rock, which was either an elevated rocky ridge or part of an island, was several miles in length and was of varying elevations.

Christine crashed into the rocky barrier, and a great wave swept Oleson up into one of the higher areas of the rocks, knocking him out. When he regained consciousness sometime later, he discovered that five of his fellow sailors had survived the crash but with significant injuries. One of the sailors had died from his wounds.

Shipwreck on a Stormy Sea by Ivan Aivazovsky

EARLY AMERICAN UFOS: 1800-1864

After resting and collecting their wits, the men realized that the rocky ridge upon which they were located did not have any trace of vegetation or animal life. They knew that starvation was going to be a very real threat. At least they would not die of thirst, as fresh rainwater was found in many of the rocky crevices. They were much revived by drinking the fresh water.

The hungry and exhausted sailors had given up hope of survival and clustered together at the base of a cliff located farther along on the ridge. Around them, the wind howled, and the furious waves continued to dash against the rocks. It seemed that their lives were near the end, when suddenly they spotted an immense flying vessel in the sky above them, being buffeted by the elements. Oleson said it was "as large as a modern (1800s) battleship," propelled by four immense wings.

The airship seemed to be heading straight toward the sailors, threatening to crash on top of them. They screamed frantically in alarm, when suddenly the airship was turned aside by the strong gust of wind, bringing it crashing down against a nearby cliff a few hundred yards from the miserable sailors. Scrambling to examine the wreckage, the sailors found a large area of debris from the crashed airship, and in the midst of the wreck were the corpses of "more than a dozen men dressed in garments of strange fashion and texture," Leander wrote. "The bodies were a bronze color, but the strangest feature of all was the immense size of the men. They ... estimated them to be more than twelve feet high. Their hair and beard were very long and as soft and silky as the hair of an infant."

The Sailors were Frightened Because in Folklore, Giants were Portrayed as Terrible Monsters (Violet Fairy Book, 1906)

Also littered among the wreckage were many strange tools, furniture, and metal boxes covered with undecipherable printed characters. Most of the tools were very large in size, making it difficult for humans to use them. Also, the crash had so badly damaged the airship that the sailors were unable to determine what means the vessel might have used to fly through the air.

Upon first observing the scene of the crashed airship, the sailors became terrified, realizing that they were looking at creatures from another world. One of Oleson's companions became so frightened by what he saw that he leaped over the

EARLY AMERICAN UFOS: 1800-1864

edge of the cliff to his death. The other men retreated from the crash site and stayed away for two days.

But extreme hunger drove the men back to the wreckage to look for anything that might be edible, preferably not the bodies of the giant humanoids. As they opened some of the boxes littering the crash site, they found portions of a very strange but "wholesome and palatable food," which saved them from starvation. After eating heartily, they summoned courage to drag the gigantic alien bodies to the cliff's edge and tumble them over. Before doing so, though, Oleson noticed what looked like a ring on one creature's giant finger and took it off.

Using materials and some of the smaller tools from the airship wreckage, Oleson and his comrades were able to build a raft, erect sails, and depart their rocky prison, now that the storm had passed, and the sea had quieted. Although they were not certain of their location, the experienced sailors headed in a direction they thought would take them to the Kerguelen Islands. However, after only six hours, they happened to run across a Russian vessel sailing for Australia, and they were rescued by the Russians. Shortly after the rescue, before they could reach port, three more of Oleson's companions passed away from their injuries and the strain of their ordeal.

Leander ended the strange narrative by saying, "Fortunately, as a partial confirmation to the truth of his story, Mr. Oleson took from one of the bodies a finger ring of immense size. It is made of

THE REAL COWBOYS & ALIENS

a compound of metals unknown to any jeweler who has ever seen it, and is set with two strange stones, the names of which are unknown to anyone who has ever examined it. The ring was taken from the thumb of the owner and measures 2 ¼ inches in diameter."

"While Mr. Oleson is an old man, he still possesses every faculty and has the highest respect for truth and veracity. Quite a number of our best citizens, among them Mr. Henry Hahn, Dr. H. C. Carleton, Green Hill and Scott Porter, saw the ring and heard the old man's story."

U.S. Census Bureau data from 1900 shows individuals in Texas named Oleson. Also, in the Census for 1900 are the names Henry Hahn, John Leander, H. C. Carleton, Green Hill, and Scott Porter. Although UFO researcher Kevin Randle has classified this story as very likely a "hoax," it is interesting that a Danish ship named *Christine* did actually exist in 1862, and that the surnames of all the persons mentioned in the story were present in the state of Texas at the time when the story was first published.

Maybe giant aliens from another world do occasionally crash land their spaceships on our planet? Though sightings of giant aliens may be rare, let us not forget the twelve-foot-tall beings spotted aboard the craft seen in the skies of Jay, Ohio in 1858. Furthermore, in 1883, a giant sword crashed into a river in New York state. But that's a story for another time and another book.

25
MEN IN BLACK IN THE CIVIL WAR
October 1863, Blakesburg, Iowa

MANY PEOPLE BELIEVE the mysterious "Men in Black," which could be humans or something else, first began appearing among us in the 1940s, but in this chapter, we will examine a strange case of a Men in Black sighting that occurred in 1863!

Today, stories of Men in Black associated with UFOs are thoroughly imbedded in the public consciousness. This is in large part thanks to the successful film franchise of the same name launched in 1997. While the films present a comedic, exaggerated take on the concept, in real life the Men in Black are believed to be government agents out to suppress the truth about UFOs. Some researchers even theorize that the Men in Black are aliens themselves.

THE REAL COWBOYS & ALIENS

Traditionally, the first real-life encounter with the Men in Black is attributed to an incident in 1947, where a witness to a UFO sighting was contacted by a government agent dressed in black. The mysterious man told the witness to keep quiet about what he saw -- or else. That is a pretty typical Man in Black encounter. No special powers, just a government agent in a black suit. But sometimes the Men in Black display paranormal attributes.

The best example of a terrifying Men in Black encounter comes courtesy of UFO researcher Albert Bender and dates back to 1952. Bender had recently formed an organization called the International Flying Saucer Bureau. Though there would be nothing odd or particularly groundbreaking about this today, remember this was not the 1970s when research into the paranormal became widespread. In the 1950s, the study of UFOs was a bit more eyebrow raising than it is today.

Shortly after forming the organization, Bender was plagued by a series of strange events, many of them demonic in nature. His attic suddenly smelled of burning sulfur, he became mysteriously ill, and he always felt as though he was being watched. Even stranger, all of this occurred in conjunction with a flap of UFO sightings in Connecticut. One night at a movie theater, he turned to see a man with glowing eyes watching him. On another occasion a man followed Bender down the street. Bender even claimed to have been hypnotized and levitated by the man!

EARLY AMERICAN UFOS: 1800-1864

*Albert Bender Standing Next to
Drawing of Man in Black (WikiWand)*

Then, one night, Bender was awakened by three shadowy strangers standing over his bed. The men possessed a letter he had mailed to a friend regarding UFOs. The men communicated with him telepathically to stop publishing his UFO newsletter. They also verbally threatened him to stop his research into the UFO phenomena in general. The three men then vanished in a cloud of yellow smoke that smelled of sulfur -- again, activity

THE REAL COWBOYS & ALIENS

more related to demons than UFO occupants. Bender immediately phoned a colleague to tell him of the terrifying ordeal. As soon as he hung up, the phone rang. When Bender answered, it was one of the three men previously in his room! He informed Bender that he would grant him that one last mistake, but he had better let the matter drop or the next mistake would be his last.

Of the three men, Bender said, "All of them were dressed in black clothes. They looked like clergymen but wore hats similar to [the] Homburg style" *(bportlibrary.org)*. Initially, Bender did shut down *Space Review*, his IFSB newsletter, concluding its run by stating, "The mystery of the flying saucers is no longer a mystery. The source is already known but any information about this is being withheld by orders from a higher source. We would like to print the full story in *Space Review* but because of the nature of the information we have been advised in the negative. We advise those engaged in saucer work to be very cautious."

But, in later years, Bender apparently lost his fear of the Men in Black and wrote a whole book about his ordeal, *Flying Saucers and the Three Men*, released in 1962.

As you can tell, the real Men in Black differ greatly from their movie counterparts. Being telepathic and having other supernatural abilities, these Men in Black may not be government agents, but some kind of aliens themselves. This was actually Bender's opinion -- that the men were not of this Earth. And for those no doubt wondering --

EARLY AMERICAN UFOS: 1800-1864

no, the Men in Black never did get Bender. He passed away at the age of 94 in 2016.

As you can no doubt guess due to the subject matter of this book, 1952 wasn't the first instance where frightening Men in Black were sighted; it was simply the first known encounter linking them to UFOs. Before this, Men -- and even Women -- in Black terrorized many people over the years in early America.

One story, set during the Civil War in southeastern Iowa, stands out in particular. But first, some background. Smack dab in the middle (geographically speaking) between the North and the South during the Civil War, the Midwest played a key role in winning said war. Not only did Midwestern men go to fight for the Union -- around 750,000 of them in fact -- but the agriculture and industry of the mid-west was invaluable to the Union and the Northern States during this trying time.

Political Cartoon Depicting Copperheads

THE REAL COWBOYS & ALIENS

During this time there was also a rather unpopular sect of the war, especially in the Midwest, called Copperheads. Though the name sounds threatening, it was actually attributed to a group of Peace Democrats. Though they refused to fight, they actually sided with the Confederacy. The Copperheads feared the social equality that would be brought about due to the end of slavery. We bring these people up because the following story concerns a group of these people being harassed by the Men in Black.

And no, we are not grasping at straws here, because we find it quite humorous that the original article, in italics, stresses that these strange men were "*dressed in black!*"

Perhaps appropriately, the article was published around Halloween in 1863 in the *Ottumwa [Iowa] Courier* with the title "Very Strange If True! -- Two Ghosts in Wapello County!" The article begins, as so many of them do, by establishing the credibility of the witness, "We learn, by a gentleman of unquestionable veracity, that great consternation prevails in Adams township, in this county, occasioned by the nightly visitations of two seeming men, at the residence of Mr. [William] Spaulding, who lives five miles east of Blakesburg. These visitors, be they who they may, and whether in the flesh or spirits in human shape, make their appearance about seven o'clock in the evening, and remain until about five in the morning, their first appearance being on Friday, a week ago."

Like the Bender encounter of the 1950s, right away these men are identified as being more like

EARLY AMERICAN UFOS: 1800-1864

specters than flesh and blood men. The paper offered the following description: "They seem medium sized, heavy set men *dressed in black*!" Again, the emphasis is not our own, but the original paper's. The article continues, "On their first appearance, on Friday night, the family and some of the neighbors, were boiling molasses about forty rods from the house, when about seven o'clock, suddenly, clubs, [corn] cobs and small sticks, began to fly in a shower, from a certain direction, occasionally hitting some one of the persons present, but generally falling in one small place."

As the assault continued, Mrs. Spaulding had a candlestick knocked from out of her hands. At the time, the attack seemed to come from out of nowhere, as the perpetrators could not be spotted. "No person was then visible, but they heard something walking about with a heavy tread," the paper reported. As it is, accounts of unseen poltergeists pelting houses with rocks were not uncommon at this time. And, again, the real-life Men in Black would seem to have as much in common with poltergeists as they do with aliens, as odd as that might sound.

To return to our tale, around 1 a.m. the family quit their boiling, and two of the men, Harrison Wellman and J. M. Spaulding "started out in the direction from whence the missiles seemed to come, armed with clubs and brickbats, to find and chastise these strange and curious intruders."

Soon the men were accosted by another volley, this one even worse. The men returned fire, throwing what they could in the direction of the objects directed at them "but without evidence of their hitting anybody."

The men returned home soon after, and as soon as they walked in the door more objects began to strike their home. The article continues, "Spaulding and one of the men went out to turn out the horses, taking their guns with them. No sooner were they out of the house than a large club fell near them, seemingly coming from behind. One of the men wheeled, and saw a man standing near enough to be distinguished in a dark night, at whom he instantly shot. The man ran and disappeared. They turned out the horses and returned to the house."

For the next four nights the family was plagued by the strange men, for on Tuesday night two of them were seen at once. The Spaulding family's neighbors also became alarmed by the phenomena and met at the Spaulding home to try and solve the mystery. The paper writes that, "On Monday night, J. W. Wellman, Wm. Hayne, Wm. Spaulding and his son, spent the night watching. Sometime in the night, one of the men looked out of the window and distinctly saw, by the light of a bright moon, a man standing before the door. After sitting awhile, they looked out again, and saw the man prostrate on a plank, lighting a dark lantern with a match. Getting their revolvers and guns ready, the party prepared to open the door, but strange to say, the four powerful men could not open it. They afterwards remained quietly in the house until 5 o'clock in the morning. One night, Mr. E. B. Day took his dog, a very sagacious animal, and tried to

EARLY AMERICAN UFOS: 1800-1864

set her on, but she trembled, ran between her master's legs, and refused to make any more demonstrations against the ghosts."

Thus ends the account of these strange men dressed in all black. And how is it that we know this family was made up of Copperheads? The final line of the article basically confirms this, even if the term "Copperhead" is not used. The reporter writes in the next to last paragraph that, "The above are the leading facts as related to us, of the most strange phenomenon. Mr. Spaulding at first attributed the persecution to political enmity..." The writer seems to imply that Spaulding was being harassed for his stance on the war. Actually, for those who like to think that the Men in Black are government men with psychic powers, perhaps this was the earliest example of them at work? In this case, rather than threatening witnesses to keep quiet about UFOs, they were merely used to harass some Copperheads in the hope that they would relocate to another area? Of course, they could have also just been normal men dressed in black to which the witnesses attributed supernatural attributes.

In any case, the article concludes, "We give the bets precisely as they are related to us, merely expressing, by way of comment, our decided conviction that no Union men, in or out of the flesh, have resorted to that mode of converting Mr. Spaulding and his political friends of the error of their political ways. The story is a strange one."

THE REAL COWBOYS & ALIENS

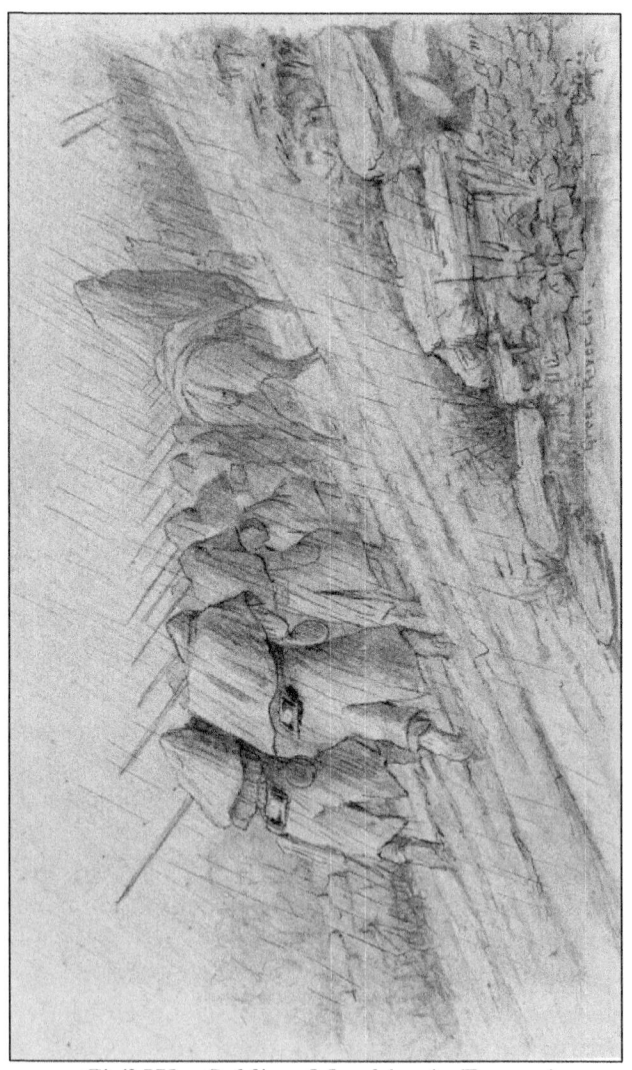

Civil War Soldiers Marching in Formation

26
ARMY FROM OUTER SPACE
September 1, 1864
Lewisburg, Virginia

IN TERMS OF THE paranormal or supernatural, the Civil War era was full of stories that at the time were called "Signs and Wonders." In the 1840s, several people once observed in the heavens a strange star formation that spelled out to them the numbers "1861," which was still 20 years in the future and was destined to be the year the Civil War started. The strange sight occurred in the skies over the wilderness near Greenfork, Indiana. Lorenzo D. Fox and several companions were on a raccoon hunt huddled around the campfire when the night sky suddenly grew darker and their hunting dogs became frightened. A streak of light flew across the sky, burning in red, green and vermillion. Making a strange maneuver of loops it

THE REAL COWBOYS & ALIENS

spelled out the numbers "1", "8", "6" and "1" in that order. At the time the men took this to mean that 1861 would be a landmark year. When the Civil War broke out years later, they finally understood their strange vision in the sky. There are other such stories that either prophesied the Civil War beforehand or strange sights relating to the war in the skies during the conflict.

The story we are covering here probably fits more into the "Signs and Wonders" category than it does UFO phenomena. But, at the same time, the story has many facets of UFO activity as well; so, we felt its inclusion was worthy of this volume. As mentioned before, the people of the nineteenth century described aerial phenomenon in terms they understood and did not have the modern vocabulary of aircraft and spaceships.

Panorama of Lewisburg, circa 1903

On September 1, 1864, in Lewisburg, Greenbrier county, Virginia, was witnessed a "remarkable atmospheric phenomenon." The sighting took place at the home of a Mrs. Pearcy, located a few miles west of Lewisburg at 3 o'clock in the afternoon. The strange scene wasn't witnessed by Mrs. Pearcy herself initially, but by her neighbor, Mr. Moses Dwyer, who was sitting on the porch for a visit along with a few other people. According to the article, "The weather was

EARLY AMERICAN UFOS: 1800-1864

quite hot and dry; not a cloud could be seen; no wind even ruffled the foliage on the surrounding trees. All things being propitious, the grand panorama began to move."

Civil War Battle North of Lewisburg, VA, 1862

Seemingly from out of nowhere, a cloudy formation "resembling cotton or smoke" in the size and shape of a series of doors came floating through the treetops on the adjacent hills south of the home. The number of cloud rolls was said to be in the thousands, as they passed rapidly through the air "in beautiful order and regularity." The cloud formations had a light green tinge on their edges. This strange sight lasted for about an hour. UFO researchers have long argued that UFOs generate cloud-like formations around themselves to hide in plain view. Additionally, these mysterious craft may also hide themselves within "real" clouds, if such clouds are in their area of

THE REAL COWBOYS & ALIENS

operations. In this 1864 sighting, the fact that the clouds had "a green tinge on their edges," suggests that something within the cloud, perhaps a UFO, was illuminating outward. Perhaps it was some type of projector beam that was responsible for what was seen next?

When the clouds finally dissipated, the show wasn't over yet. Following the unusual cloud formations, a new, equally strange sight emerged from the surrounding trees. What was witnessed next sounds a lot like a 3D holographic projection!

Onlookers saw a vast panorama suddenly unfold before their eyes -- a strange army of humanoids of various shapes and sizes, marching through the valley: "In the deep valley beneath, thousands upon thousands of [apparently] human beings [men] came in view, traveling in the same direction as the rolls, marching in good order, some thirty or forty in depth, moving rapidly, 'double quick,' and commenced ascending the sides of the almost insurmountable hills opposite, and had the stoop peculiar to men when they ascend a steep mountain. There seemed to be a great variety in the size of the men; some were very large whilst others were quite small. Their arms, legs, and heads could be distinctly seen in motion. They seemed to observe strict military discipline, and there were no stragglers to be seen."

This physical description of the men was odd. Was the writer implying that some men were giants in stature and others dwarf-like? Or were they suggesting that most of the men were normal sized accompanied by a few "little people." Obviously, a

EARLY AMERICAN UFOS: 1800-1864

better description would have been helpful. Also, though the original article has the army marching across the ground, it should be noted that later retellings changed this to have the men marching through the skies. Perhaps the view of the "projected image" [if it was a projection] changed depending on the angle of the viewer?

In terms of dress, we sadly can't say that the marching men were wearing any type of shiny space attire. Instead it was more down to earth, as they wore all white shirts and pants with wool hats. This is significant, as it implied that the strange army was neither Confederate nor Union but seemed to be neutral. Furthermore, this odd army was "without guns, swords or anything that indicated 'men of war.'" The article said that the strange men disappeared as "they came through the valley and over the steep hill, crossing the road, and finally passing out of sight, in a direction due north from those who were looking on."

As was typical of the articles of the time, the piece goes on to vouch for the reliability of the witnesses involved stating, "The gentleman who witnessed this is a man with whom you were once acquainted, Mr. Editor, and as truthful a man as we have in this county, and as little liable to be carried away by 'fanciful speculations' as any man living." The article also explains that four "respectable ladies" and "a servant girl" witnessed the strange phenomenon.

Then, two weeks later, on the 14th, "the same scene, almost identical, was seen by "eight or ten of our pickets at Bunger's Mill, and by many of the citizens in that neighborhood." The locale was only four miles away from the first sighting, and like that one, lasted for an hour.

Depiction of Civil War Battle, 1887

The above discussed article was published on the 15th of September and made the rounds across the country. Reporters then came to the area to interview the witnesses. Eventually, a slightly more detailed article was published giving more particulars of the sighting. First, it identified Mr. Moses Dwyer as an "honest and responsible farmer whose veracity is unimpeachable." And indeed, record searches by the authors confirm that not only was Dwyer a real person, but a respected citizen just as the article said. His son was even elected sheriff of Greenbrier County. It is also said that Mrs. Pearcy, who eventually came out onto the porch to watch the spectacle with her friends, was also a "lady of respectability and clear character." The article continues that all of the witnesses testify substantially to the same facts and "are perfectly willing to be sworn to the truth of the statement made to [the paper]."

EARLY AMERICAN UFOS: 1800-1864

This second article points out that the witnesses, though not lying, clearly had trouble describing the aerial phenomena preceding the soldiers. The article states: "The first thing seen was something that the witnesses do not seem able to describe with clearness and accuracy. They say it was masses or bodies of vapor, mist or something else five or six feet high and two or three feet wide, floating in a perpendicular position, above the treetops, moving on in a line with the utmost regularity and precision, then passing through the treetops, without having the line broken or disturbed. These bodies are described as being of a whitish, green color, and passed off in the distance."

As is usually the case with follow-up articles a few varying details are divulged. While in the previous article it was implied that the formation was flawless, this article gives a different take on it with one of the witnesses stating that "occasionally one [of the soldiers] would lag a little behind, and could be distinctly seem to quicken his pace to regain his position in the line... Their general appearance was white, and they were without arms or knapsacks."

The writer of the article concludes, "I put myself to some trouble to ascertain the facts and question the witnesses separately. They are above suspicion. I have given all the material facts except that the so-called men were marching north were northwest, right through the mountains. They were of all sizes, and as much like men as if they had been real flesh and blood."

More follow-up articles did their best to explain the strange phenomena as perhaps being a mirage, while others felt it was a legitimate example of "Signs and Wonders" pertaining to the war. Of course, back then, in spite of the atmospheric phenomenon no one thought to possibly classify this as any type of alien encounter, as back then there was no such idea yet in the public consciousness. Instead, both sides, North and South, considered the sighting to represent heavenly forces coming down to earth to aid their side of the battle.

THE ROCKY MOUNTAIN METEOR

September 1864
Cadotte Pass, Montana

THE ROCKY MOUNTAINS play an important role in Western lore and encompass more than 3,000 miles, spanning from Canada all the way to faraway New Mexico. The Rockies were home to many Native American tribes and to the "Mountain Men," true pioneers of the Old West who braved harsh winters and nature's various other elements to survive. Many of these men were fur trappers who dedicated themselves to catching animals like beavers so that they could sell their fur.

James Lumley was one such fur trapper in Montana, who witnessed what many people now believe was the crash landing of a UFO in the

THE REAL COWBOYS & ALIENS

Rocky Mountains. In September of 1864, Lumley was high in the mountains, trapping in an area known as Cadotte Pass. One night after the sun had gone down, he saw a bright object streak through the sky. A newspaper article that appeared later in the *St. Louis (Missouri) Democrat* and reprinted in many other newspapers described the object as a "bright, luminous body in the heavens, which was moving with a great rapidity in an easterly direction."

About five seconds after Lumley spotted it, the large object in the sky split into several smaller pieces with a flash like a "sky-rocket." Today, through the lens of Ufology, we might even go so far as to speculate that the large object was a "mothership" and that several smaller scout ships were expelled from it to the Earth below.

Be it meteor or mothership, moments later Lumley heard and felt a tremendous explosion that shook the ground below his feet. This was followed by an eerie wind that swept through the forest with a loud rushing sound. There was also the distinctive smell of sulfur in the air, similar to lit gunpowder.

The article stated, "A few minutes later he heard a heavy explosion, which jarred the earth very perceptibly, and this was shortly after followed by a rumbling sound, like a tornado sweeping through the forest. A strong wind sprang up about the same

EARLY AMERICAN UFOS: 1800-1864

time, but as suddenly subsided. The air was also filled with a peculiar odor of a sulfurous character."

The next morning, Lumley went out to investigate. What he found was astonishing. Two miles from his camp, as far as the eye could see, was a wide path of destruction. "A path had been cut through the forest, several rods wide -- giant trees uprooted or broken off near the ground -- the tops of hills shaved off, and the earth plowed up in many places. Great and widespread havoc was everywhere visible." Lumley followed the trail of destruction until finally he found the mysterious object he had seen explode the night before. Embedded in the side of a mountain was what he described as "an immense stone."

Artist Neil Riebe's Conception of 1864 Meteor

THE REAL COWBOYS & ALIENS

It was soon clear that the object was not a fallen meteor. Lumley claimed that it was "divided into compartments" and parts of it had been "carved" with hieroglyphics, similar to the writings of ancient Egypt. The newspaper said, "He [Lumley] is confident that the hieroglyphics were the work of human hands, and that the stone itself, although but a fragment of an immense body, must have been used for some purpose by animated beings." Interestingly, strange writing that looks like hieroglyphics is seen very often in UFO cases. This was one of the very first cases where such mysterious writing is said to have been seen by an observer at the scene of a crashed UFO.

Typical Egyptian Hieroglyphics (Wikipedia)

Lumley also found, littered around the crash site, fragments of glass. In addition, the ground near the crashed object was stained with some type of mysterious liquid --perhaps some kind of fuel that had leaked from the ship's engines?

The newspaper article went on to say, "Strange as this story appears, Mr. Lumley relates it with so

EARLY AMERICAN UFOS: 1800-1864

much sincerity that we are forced to accept it as true." The article also mentions several other sightings of similar objects in the same time period, including another case where a large object split into several smaller ones.[11]

The newspaper continued, "Astronomers have long held that it is probable that the heavenly bodies are inhabited -- even the comets -- and it may be that the meteors are also. Possibly, meteors could be used as a means of conveyance by the inhabitants of other planets, in exploring space, and it may be that hereafter some future Columbus, from Mercury or Uranus, may land on this planet by means of a meteoric conveyance, and take full possession thereof -- as did the Spanish navigators of the New World in 1492, and eventually drive what is known as the human race into a condition of the most abject servitude. It has always been a favorite theory with many that there must be a race superior to us, and this may at some future time be demonstrated in the manner we have indicated."

So, what is one to make of James Lumley's mysterious space stone? Interestingly, in the first major novel about an alien invasion of Earth, H. G. Wells' *The War of the Worlds* (1898), the first alien spaceship that arrives is mistaken for a meteor. Its exterior is encrusted with dirt and rocks. Only when the aliens inside it began to unscrew the lid did the humans realize that the object was an artificial cylinder.

---

[11] Likely what we refer to as the Mars Mummy, which we cover in a later chapter.

THE REAL COWBOYS & ALIENS

Even our own spacecraft, when they return to Earth, are often burned and distorted by their journey through the atmosphere. Is it possible then that what Lumley saw was a spaceship rather than a meteor? Or was it merely a very strange rock from outer space that somehow became etched with markings that looked like hieroglyphics?

The mystery remains. Below is the original story, as it appeared in many of the nation's newspapers in 1865, including the *Chicago Tribune*; the *Herald and Torch Light* of Hagerstown, Maryland; the *Weekly Republican* of Plymouth, Indiana; the *Perrysburg (Ohio) Journal*; the *St. Louis (Mo.) Democrat*; the *Vicksburg (Mississippi) Herald*; the *Brooklyn (N.Y.) Daily Eagle*, the *Charlotte (N.C.) Democrat*, and more.

Brooklyn Daily Eagle

REMARKABLE PHENOMENON.

A Piece of the Moon Found in the Rocky Mountains.

[From the St. Louis Democrat.]

Mr. James Lumley, an old Rocky Mountain trapper, who has been stopping at the Everett House for several days, makes a most remarkable statement to us, and one which if authenticated, will produce the greatest excitement in the scientific world.

Mr. Lumley states that about the middle of last September he was engaged in trapping in the mountains, about seventy-five or one hundred miles above the Great Falls of the Upper Missouri, and in the neighborhood of what is known as Cadotte Pass. Just after sunset one evening he beheld a bright luminous body in the heavens, which was moving with great rapidity in an easterly direction. It was plainly visible for at least five seconds, when it suddenly separated into particles, resembling, as Mr. Lumley describes it, the bursting of a sky-rocket in the air. A few minutes later he heard a heavy explosion, which jarred the earth very perceptibly, and this was shortly after followed by a rushing sound, like a tornado rushing through the forest. A strong wind sprang up about the same time, but suddenly subsided. The air was also filled with a peculiar odor of a sulphurous character.

These incidents would have made but slight impression on the mind of Mr. Lumley, but for the fact that the ensuing day he discovered, at a distance of about two miles from his camping place, that, so far as he could see in either direction, a path had been cut through the forest, several rods wide, giant trees uprooted or broken off near the ground, the tops of hills shaved off and the earth plowed up in many places. Great and widespread havoc was everywhere visible. Following up this track of desolation, he soon ascertained the cause of it in the shape of an immense stone that had been driven into the side of a mountain. But now comes the most remarkable part of the story.

An examination of this stone, or so much of it as was visible, showed that it had been divided into compartments, and that in various places it was carved with the curious hieroglyphics. More than this, Mr. Lumley also discovered fragments of a substance resembling glass, and here and there dark stains, as though caused by a liquid. He is confident that the hieroglyphics were the work of human hands, and that the stone itself, although but a fragment of an immense body, must have been used for some purpose by animated beings.

Strange as this story may appear, Mrs. Lumley relates it with so much sincerity that we are forced to accept it as true. It is evident that the stone he discovered was a fragment of the meteor which was visible in this section in September last. It will be remembered that it was seen in Leavenworth, in Galena, and in this city, by Colonel Rooseville. At Leavenworth it was seen to separate in particles, or explode.

Astronomers have long held that it is probable that the heavenly bodies are inhabited—even the comets—and it may be that hereafter some future Columbus, from Mercury or Uranus, may land on this planet, by means of a meteoric conveyance, and take full possession thereof—as did the Spanish navigators of the New World in 1492, and eventually drive what is known as the "human race" into a condition of the most abject servitude. It has always been a favorite theory with many that there must be a race superior to us, and this may, at some future time, be demonstrated in the way we have indicated.

The Brooklyn Daily Eagle, 11-14-1865, p.4.

28
THE BLOB FROM OUTER SPACE
October 9, 1864
Hubbardston, Massachusetts

OF ALL THE 1950s era monster movies, 1958's *The Blob* might be the most quintessential. The movie is well remembered as the first leading role for a pre-stardom Steve McQueen, who plays a hot rod loving teenager. Together he and his girlfriend, Janie, observe a strange shooting star that crashes down in the woods. An old hermit comes across the meteor and pokes it with a stick. It cracks open revealing a clear, gelatinous substance. The strange blob crawls up the stick and onto his arm. The old man is picked up by Steve and Janie and taken back into town. From there, the blob, now red from eating the old man and other victims, goes on a rampage of terror absorbing every human being it can catch.

THE REAL COWBOYS & ALIENS

The blob famously sneaks into a theater, oozing through the projector hole to terrify the audience which runs screaming out into the streets. Today, the town where that scene was filmed -- Phoenixville, Pennsylvania -- reenacts the famous scene as part of their "Blobfest," as though it were a real event. And yet, strangely enough, *The Blob* was partially inspired by a true E.T. encounter.

Ironically, the real inspiration also took place in Philadelphia, Pennsylvania, eight years before the movie was made. On September 26, 1950, two police officers, John Collins and Joe Keenan, were on their usual patrol on Vare Boulevard and 26th Avenue when suddenly, a strange purple object floated across the beam of their headlights about half a block in front of them. They watched in amazement as the strange purple "what's it" landed in an open field. Naturally, they went to the landing site to investigate. There they found a purple domed disk composed of a quivering astral jelly that measured six feet in diameter!

The shape of the object was similar to what Professor Rufus Graves found in Amherst, Massachusetts back in 1819. However, at six feet in diameter this sample was much bigger! At its highest point, the dome was one foot thick. Upon turning off their flashlights, the two policemen were shocked to see that it glowed with a strange mist. Odder still, the men felt that the strange substance was somehow alive. Not sure how to handle this odd, potentially harmful substance, they radioed for backup. They were soon joined by Sergeant Joe Cook and Patrolman James Cooper.

EARLY AMERICAN UFOS: 1800-1864

When the lawmen tried to pick up samples of the giant, Jell-O-like object, it began to fall apart. The bits that stuck to their hands began to evaporate. Thirty minutes later, the entire substance had disappeared.

The pioneer era of the 1800s had several similar incidents occur, such as the aforementioned incident in Massachusetts studied by Rufus Graves. We ended that chapter on the note that supposedly another, similar incident happened only "a few years later" in the same general vicinity. While we're uncertain as to whether or not this is the incident spoken of, as it's certainly more than just a "few years later", in October of 1864, another object crashed down in Hubbardston. As it is, Hubbardston is only about 40 miles from Amherst, where the "blob" of 1819 touched down.

The *Worcester Spy* out of Massachusetts from October of 1864 prints this story that predates both *The Blob* and the Philadelphia Incident by nearly a hundred years. It relates how a "large meteor" fell near the shore of Parker's Pond in Hubbardston on October 9th. Like Rufus Graves from 40 plus years earlier, an investigator the next morning went to the impact sight and found "a mass...of a gelatinous, light colored, semi-transparent substance, described by some parties to be as large as a hogshead." This is interesting, as that was about the size the blob from the movie was when it first popped out of the meteor.

Unfortunately, like the previous cases, the mass dissolved/evaporated rather quickly.

THE REAL COWBOYS & ALIENS

The article reported that, "A gentleman who visited the spot three days afterwards, after a large quantity had been carried away, and much more trampled into the earth, and dissolved and evaporated, says that he could at that time have gathered two bushels of the debris." The man then presented these two bushels to the Natural History Society at its regular meeting. Sadly, "Although tightly corked in a bottle, it had diminished considerably in bulk, and was partially dissolved." The paper described it as being of "a light straw color" and having "a strong odor of sulphureted hydrogen, with a sulfurous taste." This is in line both with what Rufus Graves found in 1819 and the description of the smell given by Horace Palmer, slimed by a passing meteor over Dunkirk, New York, in 1842.

The article concludes by stating that a chemical analysis would be made by a "Professor Bushee" who would then present his findings at the Natural History Society. A search of historical records encouragingly turned up a real Professor James Bushee, a curator of Minerology of the Worcester Society of Natural History. Though we were able to find notes and information on Professor Bushee's other projects and exploits, sadly, we could find nothing on the enigmatic blob sample given to him.

29
MUMMY FROM MARS
1864, North America

IN DECIDING WHETHER to include the following highly suspect story in our book, we were forced to suspend realism to some extent. Yes, we know it is highly unlikely to be a true story; however, we cannot be 100 percent certain it is not. We include it here for its historical value as a fascinating story that is almost certainly a hoax but could conceivably be true, however small that possibility is.

The yarn comes from the June 17, 1864 edition of the French newspaper *Le Pays* but the story is set in America. The tale concerns two geologists known as Paxton and Davis who find an egg-shaped rock. They crack it open only to find strange cavities inside, and within one they find a white metallic jar engraved with strange

THE REAL COWBOYS & ALIENS

hieroglyphics. Under the floor of this cavity they find another jar similar to the first, only this one has a surprise inside.

Within the container was a thirty-nine-inch mummified body "covered with a calciferous mass." The body appeared to be hairless and was otherwise human aside from a strange trunk growing from its forehead. This almost brings to mind the Pascagoula aliens of 1973 which had strange protuberances growing out of their heads. After this, the incredible discovery [never printed in America] was not heard of again until thirteen years later.

In an article for the web site *MysteriousUniverse.org*, Theo Paijmans elaborates on the French newspaper story, which he classifies as a hoax, "The hoax originated in France, in a newspaper called *Le Pays*. In its 17 June 1864 edition it published an astonishing account, in the form of a letter originating from Richmond. Its writer mentioned how a geologist ... and a few others found and inspected an immense aerolite or meteorite with a hollow chamber that contained, much to their utter amazement, an engraved plate with weird hieroglyphics but also the mummified or calcified remains of an alien being. After having deciphered the strange hieroglyphics, the team concluded that the humanoid had come from the planet Mars in some distant past. The mummified or fossilized corpse shared many of our biological features, so it followed that life on Mars resembled life on Earth. Other French

EARLY AMERICAN UFOS: 1800-1864

newspapers were quick to pick up the news under the header 'An Inhabitant of the Planet Mars.'"

"According to the letter, a man named Paxton and a geologist from Pittsburgh named Davis found, in the land of the Arapahos, in the vicinity of the James Peak, a huge meteorite with a diameter of 40 meters that had lain there 'for millions of years.' It took a team six days of digging the enormous mass out of the earth where it had crash-landed and digging into it. They found a metal vase covered with weird inscriptions. Further on they found a metal covering of about two meters that had become black with oxidation. It took them three days in removing the metal lid, after which they found a tomb, rectilinear hewn in granite and bedecked with stalagmites. In the center of the sarcophagus they found 'a man enveloped in sheets of calcium. It was the remains of a man!'"

"The humanoid measured about four feet. Face and limbs were buried beneath sedimentary sheets of calcium. The head was almost intact; no hair, smooth skin, two gaping holes where the eyes had been, very long arms, and 'five fingers of which the fourth was much shorter than the others.'"

"It was, the newspaper stated, 'an interplanetary voyager' and nothing else, 'no weapons or ornaments', were found, only a small silver plaque. On it, amongst others were drawings of the sun and the other planets. By measuring their distances, the discoverers found out that they represented the planets in our solar system, with Mars much more prominent than the others. *Le Pays* concluded that: 'Is in this distinction granted Mars to the detriment

THE REAL COWBOYS & ALIENS

of the other planets, says the perceptive correspondent, not found the pride of its inhabitant?'"

The jaw-dropping story just sounded too good to be true. According to Paijmans, the tale was traced to the fertile imagination of "a young French science editor and journalist named Henri de Parville. Henri de Parville was the penname of Francois Henri Peudefer (1838-1909). Under the name Parville he wrote many articles in scientific magazines...."

Despite widespread scorn and ridicule from the scientists of his day, Parville expanded his "mummy from Mars" narrative into a full-length book called *An Inhabitant of the Planet Mars*, published in 1865. In the book, he more fully describes what he still was trying to pass off as a real event.

EARLY AMERICAN UFOS: 1800-1864

Despite it being labeled as a hoax by many, apparently the story was too good to disappear completely. It resurfaced in an article published on October 13, 1877 in *La Capital*, an Argentina newspaper. It would seem our geologist friends, Paxton and Davis, are at it again, only this time along the Carcarana River in South America where they again find a mysterious rock. Like the last time, they find on the inside "some cavities inside the hard rock. In one of them the men saw several objects such as a white, metallic hole-ridden amphora-like jar with many hieroglyphics engraved on its surface." And also, like last time, "Under the floor of this cavity they discovered another one which contained a 39-inch (1.2 meter) tall mummified body covered with a calciferous mass."

This story, obviously a copy of the original, caused more of a stir this time around -- or, 100 years later to be exact. When someone caught wind of this story in the 1970s, it actually incited two expeditions to South America to examine the site where the alleged object was unearthed. Naturally, nothing was found as the second story was just a copy of the original article, which as we just discussed was made up. Eventually Ufologist Fabio Picasso found the earlier French story, connected the dots, and realized that the South American story was just a plagiarized version of the original.

Actually, the 1877 story inspired an 1878 sequel, also most likely made up. In this tale, A. Seraro, chemist, wrote to the *South Pacific Times* of Callao, Peru, to report that he had discovered an odd meteorite. Like Paxton and Davis, he

THE REAL COWBOYS & ALIENS

managed to dig inside of it where he too found a mummified alien body. This one was bigger though, four and a half feet tall to be exact. Instead of a jar with hieroglyphs, he found a silver plate. Somehow translating the hieroglyphs, he learned that it wasn't a meteorite at all, but an interplanetary vessel from Mars!

The *New York Times* caught wind of the story and published their own humorous opinion of it on August 17, that same year:

> **POOR PERU.**
> A Peruvian newspaper lately published a story concerning the finding of an aerolite of enormous size, in the centre of which the Peruvian scientific person who cut it open found the corpse of a man, together with a silver plate inscribed with hieroglyphics, asserting that the aerolite, the corpse, and the plate itself came from the planet Mars. This story was not only published in the leading Callao newspaper, but it was evidently regarded by the South American public as a clever and meritorious lie. The fact is a most melancholy evidence of the crude and inefficient way in which the Peruvians lie. There is not a country editor in the whole United States who would admit so weak and ridiculous an invention into his news columns. If this is

N.Y. Times, 8-17-1878, p. 4

"Undoubtedly, the Peruvians mean well, and tell the best lies they can invent. Indeed, it can be readily perceived that the heart of the inventor of the aerolite story was in the right place, and that his faults were those of the head. The truth is that the Peruvians have never been systematically taught how to lie. Very probably, if they had our educational advantages, they would lie with intelligence and effect, and it is hardly fair for us to

EARLY AMERICAN UFOS: 1800-1864

despise the Peruvians for what is their misfortune, rather than their fault."

Some researchers say that the original "mummy from Mars" story spun by Frenchman Henri de Parville is the oldest hoax in the history of UFO research, and it may well be. It ended up inspiring some great yarns that many Ufologists wish were true. Was there any basis in truth at all to the original story? It seems doubtful.

Odder still, more than a century later a real six inch "alien" mummy was discovered in South America! The naturally mummified remains were found outside of a ghost town, La Noria, in Chile's Atacama Desert around 2003. The tiny mummy was sold to Spanish businessman Ramón Navia-Osorio in 2012, who then had the strange specimen analyzed by Steven Greer shortly after.

The 1932 San Pedro Mummy

THE REAL COWBOYS & ALIENS

What was so strange about the specimen in addition to its tiny stature was its elongated skull, looking very much like an alien. The skeleton also possessed only ten ribs as opposed to twelve. Nicknamed "Ata," many Ufologists and believers in the paranormal were sure that proof of extraterrestrial life had been found at last. But, sadly, an extensive analysis showed that not only was Ata a human fetus, the remains were only around 40 years old. The anomalies, like the ten ribs, were attributed to genetic mutations.

This incident was preceded by a similar discovery in 1932. This one took place in the San Pedro Mountains of Wyoming. There, two prospectors were dynamiting their way through some thick rock trying to expose a gold vein. When the dust settled, they found they had exposed a small four foot by four-foot room, inside of which was stored a six-inch mummy.

> **MUMMIFIED PYGMY FOUND**
> LUSK, Wyo.—(U.P.)—A mummified pygmy, believed by scientists to be a progenitor of the present human race, was exhibited in Lusk recently. The mummy is owned by Homer F. Sherrill, of Crawford, Neb., and has baffled scientists in various parts of the country where it has been sent for classification. It was unearthed in a cave on a slope of one of the Peaks of Pedro mountain, near Casper, Wyo.

Article About the 1932 San Pedro Mummy

EARLY AMERICAN UFOS: 1800-1864

Because aliens were not yet a huge facet of the public consciousness, speculation was put forth that it was the mummy of one of the mythical "little people" of Native American legend. But, as with the case of Ata, scientists of today think it was just another case of a deformed, human infant. They key word here is *think*, as the mummy disappeared back in 1975, and as such no modern DNA testing was ever conducted.

INDEX

Adams, John, 103
Alan Hills Meteorite, 93
alien beings, 29, 129, 145
   as "little green men", 35, 52, 129
   as "little people", 12, 30, 33, 37, 39, 41-44, 47, 54, 128, 247
   as fairies, 49-58
   as flying humanoids, 59-68
   as giants, 174, 209-216
   as hairy hominids, 163-170
   in the form of a human, 57, 230
   mummified remains, 245-253
alien hieroglyphics, 238, 249
Allagash Abductions, 76
American Journal of Science, 96, 100, 118-119, 123
American Philosophical Society, 17, 24, 89
Amherst College, 93-101
Amherst, Massachusetts, 93-102, 242
An Inhabitant of the Planet Mars, 248
Ancient Aliens (TV series), 27
Aubeck, Chris, 171, 195
Baltimore, Maryland, 182
Bender, Allen, 218, 221
Benkleman, Nebraska, 151
Blakesburg, Iowa, 217-226
Blob, The, 14, 241-243
Book of the Damned, The, 119
Bridger, Jim, 127-128
Brownsville, Missouri, 122
Buffalo, New York, 133
Burritt College, 155
Cadotte Pass, Montana, 236-240
California Gold Rush, 9-10
Camden, Maine, 69-76
Caracas, Venezuela, 168
Carcarana River, 249
Chimney Rock, North Carolina, 59-68
Church, Frederic, 180
Civil War, 203, 207, 221, 227
Clark, Jerome, 147, 161
Clark, William, 40
Confederacy, 203, 205, 231
Copperheads (Peace Democrats), 222, 225
Cowboys and Aliens (film), 7
Crow Nation, 39, 42, 45
Danville, Pennsylvania, 181

EARLY AMERICAN UFOS: 1800-1864

de Parville, Henri, 248-251
Delaware River, 114
Dunkirk, New York, 136-140, 244
fairy rings, 50
Fire in the Sky, 152
Flying Saucers and the Three Men, 220
Fort Bridger, Wyoming, 127, 129
Fort Kearney, 194
Fort, Charles, 118
Franklin, Benjamin, 17, 23-26
Freemasons, 24-25
French-Indian War, 9, 106
Gamera, 87
Greenfork, Indiana, 227
Hall, Richard, 187
Herschel, William, 27
Hill, Barney, 74
Hill, Betty, 74
Hopkinton, New Hampshire, 54
hot-air balloons, 203-204
Houston, Texas, 210
Hubbardston, Massachusetts, 243
Illuminati, 25
Illustrated Silent Friend, embracing subjects never before scientifically discussed, The, 173
Indian Ocean, 209
International Flying Saucer Bureau, 218
James Peak, 247
Jay, Ohio, 171-178

Jefferson, Thomas, 10, 17-28, 88
Kecksburg UFO Incident, 111, 115
Kecksburg, Pennsylvania, 111
Kensington, Pennsylvania, 111-116
Koyuk, Alaska, 29-30, 37
La Mountain, John, 207
Lake Huron, 186
Lancaster, Pennsylvania, 183
Leicester, Massachusetts, 71
Leonid Meteor Shower, 117-126
Lewis and Clark Expedition, 10, 39-48, 103, 194
Lewis, Meriwether, 10, 40
Lewisburg, Virginia, 121, 228
Lincoln, Abraham, 205
Long Island, New York, 180
Louisiana Purchase, 10
Lowe, Thaddeus, 205
Lunar Society, 27
Mars, 1, 94, 172, 176, 247
Mars Mummy, 239-245
Men in Black, 14, 217-225
Mercury, 1, 172, 176, 239
Mexican-American War, 11
Miller, William, 146-148
Millerism, 146
Mount Adams, Washington, 103-110

THE REAL COWBOYS & ALIENS

Mutual UFO Network, 65
Myths and Legends of Our Own Land, 105
Natchez, Louisiana, 18
National Investigations Committee on Aerial Phenomena, 187
Nebraska City, Nebraska, 189, 199
New Orleans, Louisiana, 201
New York City, 203, 207
Newark, New Jersey, 182
Niagara Falls, 121, 124
Nome, Alaska, 29, 33
Nordic Aliens, 57
Ocean X Team, 83
Olmsted, Denison, 117
Oregon Trail, 9-10, 127
Paiute Indians, 128
Panola County, Mississippi, 140, 142, 148
Philadelphia UFO Sighting of 1950, 243
Philadelphia, Pennsylvania, 242
Phoenixville, Pennsylvania, 242
Poland, Ohio, 121, 123
Portsmouth, Virginia, 23, 86
Providence, Rhode Island, 55
Pryor Mountains, 40, 43, 46-47
Randle, Kevin, 209, 216
Rogers Rangers, 106-109
Rogers, Robert, 106
Roswell, New Mexico, 209
San Pedro Mummy, 252
Sanderson, Ivan T., 77
Sandia National Labs, 137
Saratoga (hot air balloon), 208
Schenectady, New York, 150-154
Scientific American, 155, 157, 160
Signs and Wonders, 12, 227, 234
Sioux Indians, 40, 42
Skinner, Charles M., 105
Spencer, Tennessee, 155
star rot, 93-102, 122, 123, 133-140, 243
Star Wars, 131
Texas State University, 187
unidentified flying object
 as airship, 171-173, 203-204, 208-210
 as luminous object, 17, 69, 89, 121, 157, 189, 197, 225
 as orbs of light, 54-56
 as submersible, 77
 disguised within cloud, 149, 229
 in association with meteors, 95, 113, 117, 179, 236
 in association with the moon, 141
Union Army, 207, 231
Union Army Balloon Corps, 205

EARLY AMERICAN UFOS: 1800-1864

Unnatural Phenomena: A Guide to the Bizarre Wonders of North America, 147, 161
Uranus, 239
Ute Indians, 129
Vallee, Jacques, 55, 171, 195
Venus, 1, 172, 176
Waldoboro, Maine, 74, 164-169
Walton, Travis, 152-153
War of the Worlds, The, 239
Washington, George, 24-25

Wells, H.G., 239
When UFOs Fall from the Sky, 209
wild man, 163, 167
Wilmington, North Carolina, 195-197
Wonders in the Sky: Unidentified Aerial Objects from Antiquity to Modern Times, 55, 171, 195
Worcester Society of Natural History, 244
Yoda, 129

ABOUT THE AUTHOR

Noe Torres is a recognized expert in the field of UFOs and the paranormal. He is an author, publisher, and member of the Mutual UFO Network (MUFON). He holds a Bachelor's in English and a Master's in Library Science from the University of Texas at Austin. He has written one of the most popular books about the famous Roswell Incident, titled *Ultimate Guide to the Roswell UFO Crash*, which is the top selling book among tourists visiting Roswell, New Mexico. He has also written several other well-reviewed books, including *Mexico's Roswell*, *The Other Roswell*, *Aliens in the Forest*, *Fallen Angel*, and *The Coyame Incident*.

Noe has appeared on several nationally-broadcast television shows, including season 2, episode 1 of the Travel Channel's *Mysteries of the Outdoors*, titled "Strange Attraction," which premiered in August 2017. In that show, he is interviewed extensively about unexplained mysteries in Big Bend National Park. Also, in 2017, Noe was featured in an episode titled "The Marfa Lights" for the TV series *Mysteries of the Unexplained*. In 2008, he appeared in season 1, episode 4 of the

History Channel's *UFO Hunters*, in a show called "Crash and Retrieval."

Noe has appeared several times on George Noory's famous radio show *Coast to Coast AM*, as well as on The Jeff Rense Program and may other shows. He is also in high demand as a speaker at UFO and paranormal conferences and festivals, having been a featured speaker at the 2017 International UFO Congress in Scottsdale, Arizona. He has also spoken five times at the annual Roswell UFO Conference and at many other UFO conferences throughout the United States and Mexico.

ABOUT THE AUTHOR

John LeMay was born and raised in Roswell, NM, the "UFO Capital of the World." He is the author of over a dozen books on film and western history such as *Kong Unmade: The Lost Films of Skull Island*, *Tall Tales and Half Truths of Billy the Kid*, and *Roswell USA: Towns That Celebrate UFOs, Lake Monsters, Bigfoot and Other Weirdness*. He has written for magazines such as *True West*, *Cinema Retro*, and *Mad Scientist* to name only a few. He is a Past President of the Board of Directors for the Historical Society for Southeast New Mexico.

ALSO AVAILABLE

COWBOYS & SAURIANS

DINOSAURS AND PREHISTORIC BEASTS AS SEEN BY THE PIONEERS

*TRUE TALES OF
PREHISTORIC PERIL
FROM THE PIONEER PERIOD!*

JOHN LEMAY

If you enjoy *The Real Cowboys and Aliens* series, be sure to check out the *Cowboys & Saurians* series, which examines the possibility of remnant dinosaurs alive and well during America's early days.

This is what Tobias Wayland, of the Singular Fortean Society, had to say about the book:

Cowboys & Saurians: Dinosaurs and Prehistoric Beasts as Seen by the Pioneers, written by John Le May, author of *The Real Cowboys & Aliens: UFO Encounters of the Old West,* and published in September of 2019, is a collection of accounts from the late 19th and early 20th century detailing encounters with seemingly impossible saurians.

LeMay provides a helpful introduction at the beginning of the book to those for whom the idea of remnant dinosaurs might be new, which can be skipped by experienced cryptozoologists already familiar with the subject, before launching directly into the tales of Old West-era monsters with 'The Pterodactyl of Tombstone.' New fans and experienced researchers alike will appreciate his approach to the topic, as LeMay provides a nuanced, well-researched history of the controversial story and its associated photograph—which may or may not exist.

This author isn't afraid to delve into high strangeness, either, and if there are elements thereof existent in an account, LeMay will tell you about them. From the Piasa Bird to the Van Meter Visitor to the Marfa Lights, LeMay offers a wide array of phenomena from off the beaten path. He provides a plethora of sources for each, drawing from historical newspaper accounts and the previous explorations of his fellow cryptozoologists; even peppering the book with the original art and illustrations that went to print, whenever possible.

LeMay doesn't offer a lot of definitive solutions to the mysteries presented in this book, being content to provide the reader with the tools to make their own determination. These stories are wildly entertaining, but hard facts, let alone proof, are difficult to come by, and LeMay is well aware of the struggle intrinsic to the field. To that end, there's no agenda here; he's not trying to sell you a paradigm. The only bill of goods within this book is a batch of intriguing stories, retold with excellent research.

Cowboys & Saurians is a well-researched, open-minded tour of the Old West's most fantastic tales of saurian encounters; sure to appeal to both new seekers and established cryptozoological researchers alike.

Printed in Great Britain
by Amazon